The Medieval Crossbow

The Medieval Crossbow

A Weapon Fit to Kill a King

Stuart Ellis-Gorman

Pen & Sword
MILITARY
An imprint of
Pen & Sword Books Ltd
Yorkshire – Philadelphia

First published in Great Britain in 2022 by
PEN & SWORD MILITARY
An imprint of
Pen & Sword Books Ltd
Yorkshire – Philadelphia

ISBN 978-152-678953-2

A CIP catalogue record for this book is available from the British Library

Typeset in Chennai, India
by Lapiz Digital Services.

Printed and bound by CPI UK

Pen & Sword Books Ltd incorporates the imprints of Pen & Sword
Archaeology, Atlas, Aviation, Battleground, Discovery, Family History, History, Maritime, Military, Naval, Politics, Social History, Transport, True Crime, Claymore Press, Frontline Books, Praetorian Press, Seaforth Publishing and White Owl

For a complete list of Pen & Sword titles please contact
PEN & SWORD BOOKS LTD
47 Church Street, Barnsley, South Yorkshire, S70 2AS, England
E-mail: enquiries@pen-and-sword.co.uk
Website: www.pen-and-sword.co.uk

Or

PEN AND SWORD BOOKS
1950 Lawrence Rd, Havertown, PA 19083, USA
E-mail: Uspen-and-sword@casematepublishers.com
Website: www.penandswordbooks.com

MIX
Paper from
responsible sources
FSC® C013604

Contents

Dedicated to my parents Michael and Margaret

List of Illustrations

Colour Plates

Acknowledgements

This is the book that I wished I had read when I started my PhD back in 2011. Ten years, one PhD, child, and pandemic later I wrote it myself, and I want to thank all the people who helped make it possible. First and foremost, my wife Emily who was supportive and patiently listened to me talk about crossbows more than any human should have to. My parents for all the support they have given me and my interest in history over the years. I must acknowledge, but possibly not thank, my daughter Kai who has made the experience of finding time to write even more interesting but who also provides me with motivation in her own way. I would like to thank my editor Rupert Harding for putting up with my many questions and reading the many drafts I sent, as well as the rest of the team at Pen and Sword for ensuring my text has all the polish that I alone could not give it. Thanks are also due to the Chester Beatty Library in Dublin for kindly letting me use one of my favourite surreal medieval images of the crossbow for this book. In addition, I must thank my wife again as well as Melissa Vandiver for their joint work crafting three valuable diagrams used in this book. Lastly, I would like to thank people who gave me advice, book recommendations, or let me bounce ideas off of them, including Adam Simmons, Richard Herzog and Nick Blanton as well as the whole community at AskHistorians which has been essential in keeping me interested and engaged in history over the years.

Introduction

On 25 March 1199, King Richard I of England (r. 1189–99) decided to patrol outside the walls of the Château de Châlus-Chabrol. This was a small castle in Aquitaine, a province he had inherited from his mother, and hundreds of miles from the Kingdom of England he had inherited from his father – not that Richard spent much time in England. He had been besieging the castle for some time and he may have decided to inspect the progress of his sappers in their attempts to breach the castle's defences, or perhaps he was planning an assault of the castle and wanted to see the state of the siege with his own eyes. Regardless of his motivation, Richard soon came to regret his decision. While he was on patrol one of the defenders looked over the wall, unperturbed by the shooting of Richard's own archers, and took a shot at the passing king with his crossbow. Richard was unarmoured, relying instead on a large shield carried by a retainer to protect him, and the defender's bolt lodged itself deep in his shoulder. Richard was carried back to his private tent, where his physicians attempted to remove the crossbow bolt. The first attempt snapped the bolt shaft which forced the surgeons to cut the king open and extract the metal head. This process was no doubt excruciating and badly damaged his shoulder. The wound soon became gangrenous, and Richard was bedridden. While the king lay dying in his camp his forces successfully completed the siege and put the castle's defenders to the sword. Richard died on 6 April, just over a week after he was shot. His brain and entrails were buried in the castle's chapel, his heart was sent to Rouen to join his elder brother's body, and his body was transported to Fontevraud Abbey in Anjou, where his father had been buried almost a decade earlier – his mother would join him there five years later.[1]

Remarking on his death, the French chronicler William le Breton (c.1165–c.1225) found a certain poetic irony in his fatal wounding by a crossbow. William accused Richard of introducing the sinful crossbow to France in the first place, so it was only fitting that it also relieved Europe of his presence. William was completely mistaken; Richard could take no credit for the invention of the European crossbow. The crossbow had been banned in inter-Christian warfare at the Second Lateran Council in 1139, nearly 20 years before Richard was born, and its origins were much older than that. While Richard's death is clear testimony that the ban was not very effective, the Second Lateran was a famous council and a man of the church like William would have been at least somewhat familiar with its decrees.

Why, then, would William include such an obviously false claim in his account of the king's death? The line about Richard's supposed introduction of the crossbow appears in only one of William le Breton's two accounts of Richard I's death. In William's *Gesta Philippi*, a prose chronicle about the life of King Philip II (r.1180–1223) – Richard's famous French rival – William's account was a nearly identical if abridged copy of the version of events presented in the work of Rigord (c.1150–c.1209), another French chronicler who wrote several decades before William. According to Rigord, Richard had besieged the castle because he desired a recently-discovered treasure – a golden figure of a Roman emperor – but he was shot and killed by an unknown crossbowman. Rigord did not provide much more detail around the events and while he did portray Richard as a man motivated by petty greed his version lacks William's accusations that Richard introduced the sin of the crossbow to France.

It was in his *Philippide*, a panegyric written to celebrate Philip II's victory at Bouvines in 1214, where William really let his imagination run wild. In this telling the death of Richard I is a 200-line literary set piece that marks the thrilling conclusion of Book V. The passage about the introduction of the crossbow is part of a 31-line speech delivered by one of the three Fates who decided that even though her sisters were still weaving Richard's life, it must be cut short. She guided Archard of Chalus – the lord of the castle and person who we are told found the treasure in the first place – to discover a hidden crossbow bolt because: 'This is how I want Richard to die, for it was he who first introduced the crossbow into France. Now let him suffer the fate he dealt out to others.'[2] Richard is then shot and killed with that same bolt, his fate justified by his lifetime of villainy.

This speech must be seen within the broader context of the work – the *Philippide* was meant to praise Richard's long-time rival Philip II and it frequently and vehemently condemns the English king in no uncertain terms. We are told that Richard I was killed because of his greed in demanding the treasure for himself despite having no right to it, and that he had no respect for God, broke treaties and violated holy days. No crime is beneath Richard in this work, and so the suggestion that he was responsible for introducing the crossbow is just another exaggerated crime of the English king.

William and Rigord were hardly the only contemporary chroniclers to remark upon the unexpected death of Richard I at a remote castle within his own realm. Numerous accounts of his death have survived, and while they often disagree with each other in terms of specific details they can all shed some light on the final hours of the crossbow's most famous victim. They also show how difficult it can be to piece together a true and accurate account of an event that happened over 800 years ago.

The English chronicler Roger of Howden (died c.1201) wrote what is probably the best-known account of Richard's death, including a famous description of the king's confrontation with his killer. Roger was an English chronicler probably best known for accompanying Richard on the Third Crusade and writing a detailed account of the expedition. On the king's death, Roger wrote that Richard was outside Chalus Castle preparing for an imminent assault when he was shot by a crossbow. Upon being shot Richard rode back to camp and told the captain of his mercenaries to begin the assault without him. Roger claimed that Richard was shot by a man named Bertrannus de Gurdon. When Richard learned that he would not survive he had Gurdon called before him – the castle having fallen by this stage and its defenders captured. Richard asked him: 'What wrong have I done to you that you should kill me?' To which Gurdon responded: 'You killed my father and my two brothers, and you wished to kill me. Take what vengeance you like. So long as you die, I shall willingly suffer any torments you may devise.'[3] Roger says that Richard forgave Gurdon and ordered him be released, but upon the king's death the captain of his mercenaries, a man named Mercadier, had Gurdon captured and flayed alive.[4]

The historian John Gillingham has suggested there are reasons to doubt Roger's account of events, however. While Roger's accounts are generally seen as accurate since in his younger years he had been intimately involved in Anglo-French politics, by the late 1190s he had retired to Howden in Yorkshire and seems to have primarily concerned himself with regional matters in and around northern England. As such, he probably was not as well informed about events in central France during this period. This means that despite Roger writing his account very soon after Richard's death – he only outlived Richard by a few years – Gillingham has argued that the Bertrannus de Gurdon story should be treated as a myth rather than fact.[5]

Another reason for doubting Roger's account is that another contemporary, Bernard Itier (1163–1225), credited Richard's death to a completely different crossbowman. While English by birth, Bernard was based in the great abbey of St Martial in Limoges from 1199 and was librarian there from 1204 until his death in 1225. This placed him much closer to the scene of Richard's death than Roger, even if he only took up residence there the same year that Richard died. Chalus Castle is only a few miles from Limoges and Bernard as librarian would have had access to all sorts of regional chronicles, annals and other accounts to bolster his own knowledge.

Somewhat confusingly, Bernard made only passing mention of Richard's death at Chalus in his entry for 1199 in the *St Martial Chronicle*, the abbey's primary chronicle that he was charged with updating during his life. This version only included the name of the king along with a series of others who had died that year. The list does not even definitively identify the dead man as King Richard, simply listing his given name and leaving it to the reader to assume the famous Richard is the one that was meant. However, Bernard was also in the habit of adding notes to other books in his collection and it is in an addendum to a copy of a chronicle by Geoffrey de Vigeois (fl. twelfth century) that we find Bernard's more detailed account of Richard's death. In this account, in which

Bernard identifies himself as the author, he provided a standard description of how Richard I was shot by a crossbow at Chalus Castle and died later of infection. However, Bernard named the crossbowman who killed the king as one Peter Basil. Unfortunately, Bernard did not provide any further detail about Peter Basil or his eventual fate but given Bernard's close proximity to the location of Richard's death he was probably better informed about King Richard's killer than Roger was.[6]

Ralph of Coggeshall (died after 1227) was an English chronicler who wrote the most detailed contemporary version of the siege of Chalus and the death of King Richard I. Ralph's version was probably written before 1202, placing it very close to the actual event. Ralph described what time of day it was when Richard was shot – after lunch – as well as many specific details about the castle and the siege in general but he made no mention of a dramatic confrontation between Richard and his killer. Ralph included no details about the identity of the archer who shot Richard. He even claimed that the defenders did not even know Richard was personally among the besieging forces and had no reason to expect that the man they shot was the king.[7]

Ralph described in gruesome detail how some of the iron from the bolt got embedded in Richard's shoulder and his surgeons bled him to try and treat his injury. The surgeons then tried to remove the last bits of iron but were unable to and the wound became infected and gangrenous. Ralph described in some detail the various treatments Richard's doctor tried to battle an infection they had no way of curing. Richard took his final confession at the hands of Milo, abbot of Le Pin, who also delivered Extreme Unction and closed the dead king's eyes and mouth. Ralph's version of events best captured the horror of watching the king die slowly and painfully while medieval medicine could do nothing to save him. It is the most emotional and intimate account of Richard's death.[8]

What do we know of the weapon that killed Richard I? The crossbow was one of medieval Europe's most iconic weapons, but it is also something of a mystery – even to people at the time. Take, for example, a document detailing the requirements for the militia of Florence written in 1260. This included detailed descriptions of the type of equipment that militia members were expected to show up to muster with as well as the fine they would have to pay for each item that was missing. The requirements dictated that an infantryman could face several distinct fines if he showed up without each of his breastplate, helmet, gorget, lance and shield. Similar requirements are outlined for the cavalry. However, when it comes to crossbowmen the document simply says:

> *Item*, all crossbowmen and archers of the city and county of Florence ought to own and are obliged to carry or have in the present army all of the arms that are required and necessary for them or suffer whatever penalty the Podesta [that is, the chief magistrate of Florence] wishes to give them.[9]

The document offers no breakdown of what equipment a crossbowman was expected to bring with him. Given how detailed the rest of the document is, we are left to speculate that either the author was not entirely familiar with the equipment of a crossbowman or, less likely I would argue, they felt that it was so obvious what that equipment was that they need not detail it. The author assumed that at minimum the Podesta would know what equipment was required since they were given power to fine negligent archers, but the vague wording clearly sets the crossbowmen apart from the rest of the militia even at a time when the crossbow was a widely used and popular weapon. The crossbow could not even claim to be a new weapon: by the time this document was written the crossbow was already well over a thousand years old and had been present in Europe for at least several centuries, if not longer.[10]

One of the earliest references to the crossbow is in the famous military strategist Sun Tzu's (c.544 BC–c.496 BC) *The Art of War*, and even that reference implies that it was already a well-established technology at the time of writing. Given the origin of this reference it should come as no surprise that the crossbow is generally presumed to have been invented in ancient China. While it was used extensively in Chinese warfare for centuries, its first appearance in Europe seems to have come with the writings of the Roman military strategist Vegetius (fl. fourth century AD). Even then it failed to emerge into the main narrative of European history, drifting along in the background until the tenth century, from when it became increasingly prominent in medieval warfare. It was the dominant weapon in much of Europe until its eventual decline during the sixteenth century – replaced as so many other weapons were by another Chinese technology: gunpowder.

The earliest Chinese crossbows consisted of a standard composite bow, the same kind that a regular archer would have used, mounted horizontally on a stock with a trigger made of bronze at the opposite end. The archer would rest his feet on the bow and pull the string back the full length of the stock to rest in the trigger. Little is known about the early medieval European crossbows, but the oldest available evidence suggests that they were mostly made with wooden bows. Unlike the Chinese crossbows, these were much thicker and shorter than those used by traditional archers and were drawn back to a trigger located approximately halfway down the crossbow stock. From around the twelfth century Europe began constructing composite bows for use with crossbows, although again these were much thicker and shorter than their equivalent hand bows. In the fourteenth century a third bow type was added to the mix with the invention of the steel crossbow. All three types of bows continued to be popular throughout the Middle Ages and the steel bow continued to thrive in the early modern era.

While the earliest European crossbows were spanned much like their Chinese forebears, with the archer putting his feet on the bow and pulling the string back, over the course of the Middle Ages a series of inventions were developed to aid the archer in drawing his weapon. The addition of a stirrup to the front of the crossbow prevented the archer from having to stand on his weapon to draw it, and the belt hook let him use his legs to span the weapon instead of his arms. The *cranequin* and windlass let the medieval archer span crossbows that were far too powerful for human strength alone, greatly expanding the varieties of crossbows that could be made. As with the bows, earlier spanning devices coexisted alongside later innovations: there was no linear progression of new technology eclipsing older forms.

That is a rather abbreviated history of the crossbow, and, as the following chapters will show, the full story is much more complicated. There are twists and turns and plenty of questions we do not have the answers to in the story of the European crossbow. There are also plenty more stories of individuals being shot at, wounded and killed by the crossbow – Richard I was not even the only example of regicide by crossbow. This book will not be the final word on the subject, far from it, instead this is an introduction to the world of the crossbow and its many facets and complications.

Part I:
The Crossbow – A Technical Study

Chapter 1

What is a Crossbow?

Writing while in exile in the Theotokos Kecharitomene Monastery in northern Constantinople in or around 1148 AD, the Byzantine princess Anna Komnene (1083–c.1150) wrote her most famous work, *The Alexiad.* This book was a biography of her father Emperor Alexios I Komnenos (r.1081–1118), founder of the Komnenian dynasty and one of the empire's most famous medieval leaders. The book covers the whole of Alexios' life but the section that has generally garnered the most attention has been the narrative around the First Crusade (1095–9). Anna Komnene, aged 13 at the time, was an eyewitness to the arrival of the Crusaders at Constantinople in late 1096 and early 1097.

Her whole account of the Crusaders is fascinating but there is one passage that is of particular interest to students of the medieval crossbow. Anna described a conflict between some Crusaders and Byzantines near Constantinople where a Byzantine soldier named Marianus had his helmet shot off his head by a crossbow and during her narrative Anna wrote the following commentary on that weapon:

> The cross-bow is a weapon of the barbarians, absolutely unknown to the Greeks. In order to stretch it one does not pull the string with the right hand while pushing the bow with the left away from the body; this instrument of war, which fires weapons to an enormous distances, has to be stretched by lying almost on one's back; each foot is pressed forcibly against the half-circles of the bow and the two hands tug at the bow, pulling with all one's strength towards the body. At the mid-point of the string is a groove, shaped like a cylinder cut in half and fitted to the string itself; it is about the length of a fair-sized arrow, extending from the string to the center of the bow. Along this groove arrows of all kinds are fired. They are very short, but extremely thick with a heavy iron tip. In the firing the string exerts tremendous violence and force, so that the missiles wherever they strike do not rebound; in fact they transfix a shield, but through a heavy iron breastplate, and resume their flight on the far side, so irresistible and violent is the discharge. An arrow of this type has been known to make its way right through a bronze statue, and when fired at the wall of the very great town its point either protruded from the inner side or buried itself in the wall and disappeared altogether. Such is the cross-bow, a truly diabolical machine. The unfortunate man who is struck by it dies without feeling the blow; however strong the impact he knows nothing of it.[1]

This passage has been the subject of great debate and confusion. Anna clearly stated that the crossbow was unknown in Byzantium, but from other sources we know that it was probably present in some capacity in Imperial Rome. Was it forgotten in the surviving Eastern Roman Empire, or was Anna either mistaken or misleading her readers? Anna was writing this account some 50 years after the event and was also writing to praise her father, and in this specific part condemning the Crusaders. It is possible that this passage was meant to denounce the 'barbaric' Crusaders with their weapon unknown in 'civilized' Byzantium. It is also possible that Anna was correct that the crossbow had fallen out of use but also used that to make a point decrying the barbarity of the Franks – the two need not be mutually exclusive.[2]

Regardless of her motive, Anna captured a fascinating moment in the crossbow's history. She wrote a clear explanation of what the eleventh-century European crossbow looked like, how it was operated, and how it performed in combat, although on this final point she clearly engaged in some exaggeration. Anna's' description does a far better job of conjuring the image of a medieval archer using his crossbow than just about anything written since.

The Crossbow

European crossbows came in several different types and changed significantly between when Anna wrote her description and the end of the Middle Ages. New kinds of bows were invented, and a vast array of devices was developed to aid in the spanning of crossbows. This latter development enabled the construction of crossbows that were too powerful to be spanned by human strength alone. To fully understand the medieval European crossbow, it is necessary to first examine its component parts and overall construction. In addition, a basic

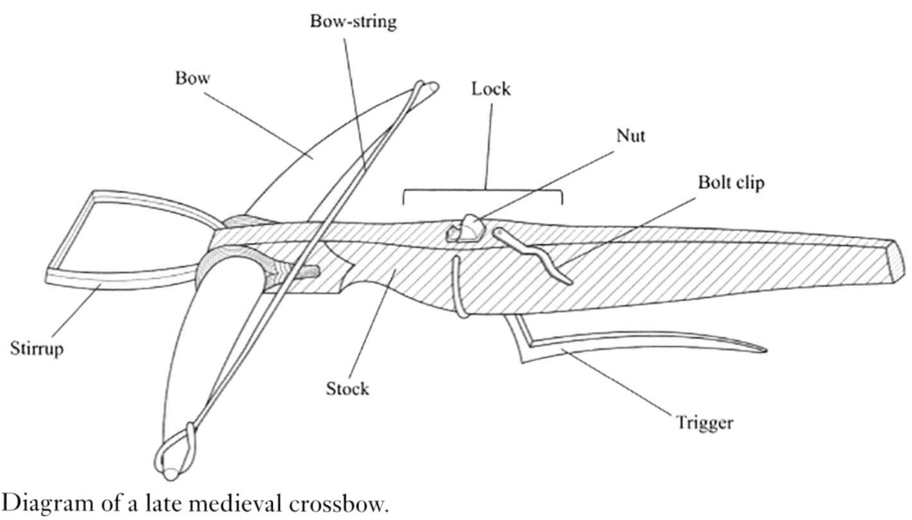

Diagram of a late medieval crossbow.

understanding of the mechanics of archery and the various methods of spanning the crossbow will benefit anyone interested in understanding its history.

The crossbow's bow

Crossbows are generally classified by the materials used to construct their bow. During the Middle Ages there were three types of bow material for crossbows: wood, composite and steel.

The oldest known crossbows had bows made of wood. So far as we can tell, the most common type of wood was yew – the same wood used for English longbows. While they used the same material, the bow for a crossbow was much shorter and thicker than a longbow. Only a handful of wooden crossbows have survived from the Middle Ages, but those that have were on average between 950mm and 1250mm in length. In contrast the average length of a longbow from the *Mary Rose* is 1980mm – nearly double the length. Despite their short length the wooden crossbow was very powerful. The Berkhamsted Bow, a wooden crossbow that was found in an archaeological dig in the mid-twentieth century, has an estimated draw weight of 150lbs, comparable to the *Mary Rose* bows from several centuries later, and some recent reconstructions have suggested that this might be an underestimate.[3]

Composite crossbows, sometimes also called horn bows, were made from a combination of wood, animal horn and sinew. A traditional handheld composite bow was made with a wooden core, a layer of horn along the belly of the bow, and with strips of sinew applied liberally along the back. The belly of the bow is the part that faces the archer, the back is the part that faces away. Composite crossbows used the same elements but varied significantly in their construction. Most still had a core of wood but the layering of the horn and wood could vary substantially between crossbows. Some had layers of horn on either side of the wood, others stacked the layers vertically, and sometimes they eschewed wood altogether. For an example of the insides of a composite crossbow now in the Metropolitan Museum of Art in New York, see Plate 1. Historians' understanding of the underlying principles of composite crossbow design are limited since it is hard to see the internal design of these weapons without breaking them open. Crossbows that are already broken can be examined and some museums have cut open a composite bow to examine the insides, but in most cases the bows are too valuable to be destroyed and historians do not have ready access to scanning machinery that might capture a picture of what is inside the bows.[4]

Composite crossbow design seems to have changed significantly in the transition from the fourteenth to the fifteenth centuries, but then remained consistent in size and design through the early sixteenth century, after which they seem to have fallen out of use. However, this is based on a fairly limited number of surviving archaeological examples, especially before the fifteenth century, so it may be that the available evidence creates the illusion of a disparity that was much smaller at the time.[5]

Composite crossbows first appeared in Europe during the twelfth century. The first definitive reference to a composite crossbow was in a book written for Saladin (1137–93) by the Muslim engineer Mardi ibn Ali al-Tarsusi (fl.1180s). This has led some scholars to believe the composite crossbow was a Muslim invention. The possibility of the composite crossbow coming from the Middle East to Europe is increased by the work of Richard I and his brother King John (r.1199–1216) to establish a crossbow industry in England. Richard's attempt floundered but in 1204 John restarted the industry and recruited expert crossbow makers from abroad, most notably a man named Peter the Saracen. Peter was paid 9*d.* a day, which was 50 per cent more than any contemporary crossbow maker working in England. Peter was also lavished with royal gifts, some of which could be as valuable as a whole month's wages. A few years later Benedict the Moor was added to the list of royal crossbow makers at a slightly lower but still premium wage of between 6*d.* and 7*d.* a day. Given their names it is very likely that both men were either Muslims, recent converts to Christianity, or from a Muslim-ruled part of the world.

The timing of their arrival coinciding with an increase in the available evidence for composite crossbows has tempted many historians to link the two. We must be cautious, however, because even though it does seem likely that Muslim craftsmen may have had advanced knowledge of crossbow making it is very unlikely that al-Tarsusi actually invented the composite crossbow. Instead, al-Tarsusi was probably describing a technology already in use and it is very probable that the composite crossbow was already in use in Europe when he was writing.[6]

Arguably the most significant development in the history of crossbows was the invention of the steel bow. Steel crossbows quickly rose in popularity during the fifteenth century and completely eclipsed all other types during the sixteenth century. This rise in popularity has led historians like Sir Ralph Payne-Gallwey to argue that the steel bow was the perfected version of the weapon.[7]

The origins of the steel crossbow is a subject of some dispute. Josef Alm was vague about the exact date of invention but argued that steel crossbows first came into common use in the mid-fifteenth century, noting a reference to them in Burgundian accounts in 1437. The oldest surviving steel crossbows also date to the fifteenth century, which would support this approximate timeline.[8] In contrast, Payne-Gallwey and Bert Hall have argued that steel crossbows first saw use in the fourteenth century. Payne-Gallwey argued that they were introduced shortly after the Battle of Crécy, while Hall placed it sometime during the late fourteenth century, but neither provided specific evidence for these dates.[9] More recently, Dirk Breiding pointed to a reference to a gilded steel bow among the items stolen from Mathilda of Brabant in 1316 as evidence of the steel crossbow's fourteenth-century origins.[10] In marked contrast with these opinions, Paul Chevedden has argued that a reference to a type of siege *ballista*, called a 'thunderbolt' *ballista*, in the writings of the anonymous Roman author of the

De Rebus Bellicis could be a reference to a siege crossbow with a heavy steel bow.[11] This interpretation has not generally been accepted and no steel crossbow survives from such an early date, but it is worth considering the possibility that the steel crossbow was a much older technology that simply became more common in the Later Middle Ages. The most likely explanation is that the steel crossbow was invented sometime during the late thirteenth or early fourteenth century but only entered widespread use in the fifteenth century.[12]

The steel bow did not replace or even eclipse the other types of bows. The Teutonic Knights continued to prefer the composite crossbow and the Burgundians made purchases of yew crossbows in the years 1433, 1437, 1442 and 1448.[13] William Ross, Master of the King's Ordnance and Controller of the Ordnance for Edward IV's (r.1461–70, 1471–83) invasion of France in 1475, bought over a hundred crossbows between the years 1473 and 1483, including both wood and steel ones. The two kinds of crossbow even had comparable costs. In 1476 there were 121 steel crossbows and 40 wooden crossbows in the wardrobe in Calais, along with 1,076 longbows. Instead of being the perfection of the weapon, the steel crossbow was just another tool added to the mix of European warfare.[14]

The stock

After the bow, the most important component of the crossbow was the stock – sometimes also known as the tiller. This piece of wood held the crossbow together and made an overpowered bow into a weapon. The stock was also often the most decorated part of the crossbow, and many ornate examples survive from the Middle Ages. For all its importance, little can be said about the stock of the medieval crossbow as a piece of design. It did not change much during the Middle Ages with only a few variations in length and thickness to be found during the period. The stock lacks the complexity of engineering found in the bow or the trigger.

The most common method for mounting the bow to the stock was by passing it through a hole near the front. An alternative method was to leave the end of the stock open and slide the bow into it from the front. To ensure a tight fit the bow was usually secured in place with wedges of wood. The bow was mounted at a slight upwards angle so that the string would pass cleanly along the top of the stock. The bow was then further secured by lashing it to the stock with two loops, one on either side of the stock. Initially these lashings were made from leather thongs, sinew or string but towards the end of the Middle Ages metal clamps, sometimes also called bridles or harnesses, were used to secure steel bows to their stocks.[15]

The design of a stock can sometimes provide insight into the types of spanning devices that were used with that particular crossbow. For example, a crossbow spanned with a windlass required a long stock because the crossbow had to rest with its front on the ground while the windlass handles had to be at a height for the archer to turn them comfortably. Similarly, the *cranequin* required a narrow stock, so that the loop that held it in place could fit around it securely, as well as a metal lug through the stock to hold it in place while it was being cranked. Stocks for use with *cranequins* tended to be shorter since they did not have the same length requirements as the windlass. Devices like the *krihake* and goats-foot lever required hooks or lugs on the tiller where they could be anchored. Attempts to identify specific spanning mechanisms from the stock can be misleading, however, as some crossbows had a long life and were sometimes modified to be used with a different spanning device. It was not uncommon to cut longer crossbow stocks short for use with a *cranequin*, for example, or to add a lug to a stock where once there was none.[16]

The crossbow nut

The crossbow nut was the piece of the crossbow that held the string in place when the weapon was spanned. It sat in a section of the crossbow known as the lock and was an essential component of the trigger mechanism. Nuts were often left to sit freely in the lock, but they could also be secured by lashings or a metal pin. It is possible that the secured nuts were a later invention, but both secured and unsecured nuts were used contemporaneously. Crossbow nuts came in both single and double hook types, based on the number of clawed hooks carved into the nut that stuck up and held the string in place. In general, double-hook nuts appear to have been the dominant style. For an example of a surviving but damaged double-hook medieval crossbow nut, see Plate 2.

While crossbow nut design did not change much during the Middle Ages there were some slight improvements over the centuries. Initially nuts had perfectly round bottoms that the trigger pressed up against but in later examples the crossbow maker carved a notch into the bottom that let the trigger rest more securely when the crossbow was loaded. This slight change improved the smoothness of shooting, which was important since early crossbows suffered from a stiff release when shot which hindered accuracy. It also ensured a better hold for the trigger on the nut which would have prevented misfires.[17]

Since nuts were usually made of antler or other very durable materials, they are the piece of the crossbow that is most likely to survive to the modern era. Even when the rest of the crossbow has rotted away, the nut can still sometimes be found by archaeologists. The presence of crossbow nuts at an excavation site makes a strong case for the existence of crossbows in or around that location. The crossbow nut is useful archaeological data when studying the broader trends in how the crossbow was used and where it was made. Surviving whole crossbows have been moved several times between their original construction and where they are stored now. In contrast the discovery of crossbow nuts often indicates places where crossbows were left or discarded, possibly very close to where they would have been used. A group of broken crossbow nuts can also suggest a place of manufacture for crossbows, since a nut which broke while it was being made would simply be discarded nearby. Annette Holts Booth used evidence of broken crossbow nuts to make the case that crossbows were probably manufactured in the archbishop's palace in Trondheim, Norway.[18]

Triggers

The most common medieval European crossbow trigger was the rolling nut trigger. The nuts have been discussed above, what held the nut in place was a trigger which resembled an elongated letter Z. The top of this Z-trigger pressed up against the bottom of the crossbow nut preventing it from spinning while the lower end hung beneath the crossbow. To shoot the crossbow the archer pulled up on one end of the trigger which operated as a lever to release the pressure on the nut allowing it to turn under the force of the bow. Medieval crossbows had no mechanism for the trigger returning to its original position and the nut had to be manually returned to its upright position. Later designs included a small metal spring inside the stock that would press down on the trigger to ensure that it always applied pressure to the nut.[19]

The rolling nut was a relatively simple and reliable trigger system. The problem it suffered was that it could require substantial force to pull the trigger. The pressure exerted on the nut by the string could make it difficult to dislodge the trigger from it, especially when the trigger was just a simple lever. This hard trigger pull could impair accuracy because it shifted concentration away from aiming, and the force of the string release could cause the crossbow to jerk suddenly.

These difficulties led to an innovation. A simple mechanism called a sear was inserted into the middle of the trigger process so that the amount of force applied to the trigger was amplified, meaning it required less force overall to shoot the crossbow. This trigger had extra complexity in that it required an extra re-arming process of the middle mechanism before it could be shot again. To reload the weapon the archer had to pull a string that hung from the middle

Diagram of a fifteenth-century crossbow rolling nut trigger with a spring, after Josef Alm.

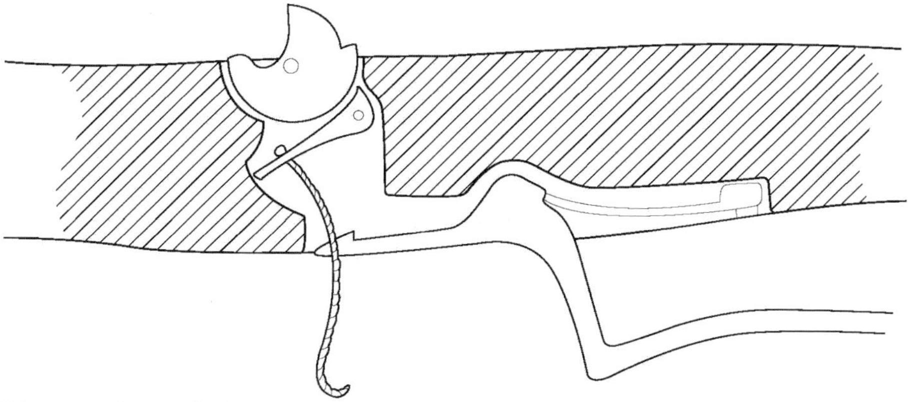

Diagram of an early sixteenth-century crossbow trigger with a sear, after Josef Alm.

of the tiller which would lift the mechanism back to its starting location. This new trigger system only came into use towards the end of the Middle Ages.[20]

One of the advantages of the standard Z-trigger is that it was simple and sturdy. The trigger was very unlikely to break and while the nut was more fragile it was also easily replaceable. This new trigger had an extra set of moving parts, which presented an extra opportunity for something to go wrong and render the weapon useless. Further innovations in crossbow-trigger design continued in the early modern era with triggers becoming increasingly complex from the sixteenth century on. These were generally limited to crossbows for hunting and target shooting – areas where a malfunction was less likely to be fatal.[21]

While the rolling nut was the most common form of trigger in medieval Europe, it was not the only one. Rarer but still attested to in the surviving evidence was the notch and pin trigger, sometimes also known as the peg-in-hole trigger. For this trigger a notch was carved into the stock and when spanned the string would rest within the notch. When the trigger was compressed, a peg positioned beneath the notch would push the string up. An unusual variant of this trigger was found in 1972 when a crossbow possibly dating to the eleventh century was excavated near Grenoble in France. This crossbow used the same principles of the notch and pin trigger but instead of pushing up from beneath the trigger was on the top of the stock and when pushed down it would pry the string up from the notch. One advantage of the notch and pin trigger was that it could be made entirely out of wood: there was no need for a horn nut or metal Z-trigger. This has suggested to some historians that it may be one of the oldest forms of crossbow trigger, but with the limited evidence available on early crossbows this is still largely speculative.[22]

Spanning Devices

Initially crossbows were spanned by hand in the manner described by Anna Komnene at the start of this chapter. Over the course of the Middle Ages a series of spanning devices were developed that made it easier for archers to draw their weapons and, in some cases, allowed them to span weapons beyond the limits of human strength alone. The crossbow, especially in the Later Middle Ages, was not a single technology but rather a whole range of similar weapons with subtle but important differences in power and design. A variety of spanning devices were developed in conjunction with these crossbows. Just as it would be foolish to discuss guns without covering gunpowder, no discussion of the medieval crossbow is truly complete without some discussion of the various machines used to assist in their spanning.

Stirrups and belt hooks

The earliest method to assist with spanning a crossbow was the addition of a stirrup to the front of the crossbow. This gave the archer a more stable position to place their foot when spanning the weapon. The previous practice of placing the archer's feet on the bow itself had created significant balance issues so many archers spanned their weapons by sitting or lying down on the ground. Having to sit down and stand up between shots was both slow and exhausting, so the stirrup, while somewhat mundane, definitely made the practice of spanning a crossbow much easier. It also prevented any damage to the bow that might result from pressing the archer's feet directly against it. The stirrup was probably introduced sometime during the twelfth century and by the thirteenth century nearly all crossbows had stirrups. Jean de Joinville made a clear reference to the use of stirrups in crossbows in his account of the Seventh Crusade (1248–54), the casual nature of his account implying that the stirrup was already a common feature by then.[23]

An archer spanning a crossbow with a belt hook – from Ralph Payne-Gallwey, *The Crossbow: Its Military and Sporting History, Construction and Use* (London, 1903).

The belt hook emerged at around the same time as the stirrup and was a complementary technology. The belt hook was, as it sounds, a metal hook hung from a belt around the archer's waist and came in both single and dual clawed varieties. When using a belt hook the archer put the string in the claw and their foot in the stirrup. After that there were two methods of spanning the crossbow. They could either balance on one foot and push down on the stirrup to pull the string into place, or they could crouch down on one knee and push themselves up, using both leg and back muscles, to span the crossbow. There are contemporary illustrations that show both methods and it is not clear which, if either, was more popular. The belt hook was introduced in the twelfth century and was in common use for the rest of the Middle Ages. The advantage of the belt hook over just the stirrup alone was that it allowed the archer to use their more powerful leg muscles to span the bow rather than their arms or back muscles alone. The hook also prevented the archer from having to pull back on the string with their fingers, an experience that would have been increasingly painful as crossbows became more powerful.[24]

The windlass

The windlass was a system of ropes and pulleys connected to a winch (Plate 3) which enabled an archer to span a crossbow that was beyond the limits of human strength alone. When in use the windlass was mounted on the end of the crossbow's stock and a set of hooks were attached to the string. The archer would then spin the handles to pull back the string and span the crossbow. Windlasses were very powerful machines, but they were also slow and cumbersome. The archer had to be careful not to tangle the windlass's ropes, and the process of setting up the windlass and fully spanning the crossbow could take between 30 seconds to a minute depending on the power of the crossbow and the experience of the archer using it.[25]

The earliest definite mention of the windlass comes, once again, from the book on weapons written by al-Tarsusi, discussed earlier in this chapter. The introduction of the windlass to Europe has often been linked to Richard I's return from the Third Crusade, but the early evidence for the windlass is vague and it is entirely possible it was already present in Europe earlier in the twelfth century. The first possible reference to the windlass in England actually dates to the reign of Richard's brother John and the first clear evidence is from the reign of Henry III (r.1216–72). The windlass remained a very popular spanning device well into the modern era, most especially with Flemish target shooters.[26]

The cranequin

A second spanning machine that was created for use on high-powered crossbows was the *cranequin*. It was secured to the top of the crossbow with a loop that wrapped around the stock and was usually prevented from slipping too far down the crossbow by a lug through the stock. The *cranequin* was essentially a gearbox and shaft with a crank handle to power it (Plate 24). One end of the gear stick had a pair of hooks that were attached to the string so that when the handle was cranked the string would slowly be pulled back notch by notch. The *cranequin* was probably invented in the fourteenth century and continued to be used well into the modern era.[27]

The *cranequin* is the spanning device that survives in the greatest number from the Middle Ages which means it is also one of the best understood by historians. There were two primary styles of medieval *cranequin*: German, which had the crank handle mounted on the top, and Spanish, which had the crank mounted on the side. These labels described the origin of the aesthetic and not necessarily their place of manufacture: other regions could make a *cranequin* in the Spanish style, for example.[28]

As devices made entirely of metal, the weight of the *cranequin* would have been of some concern to the archer using it. Most *cranequins* probably weighed a little over 2kg, but surviving examples weigh as little as 1kg or as much as 5kg. This may not seem like a lot at first glance, but a *cranequin* could potentially increase the weight of the crossbow it was being used with by at least 50 per cent. This means that, despite their compact size, when using a *cranequin* the archer often braced the weapon against something, either their body or the ground, especially if they planned to be shooting for a while. In fact, numerous medieval depictions of archers using a *cranequin* show them crouching in a

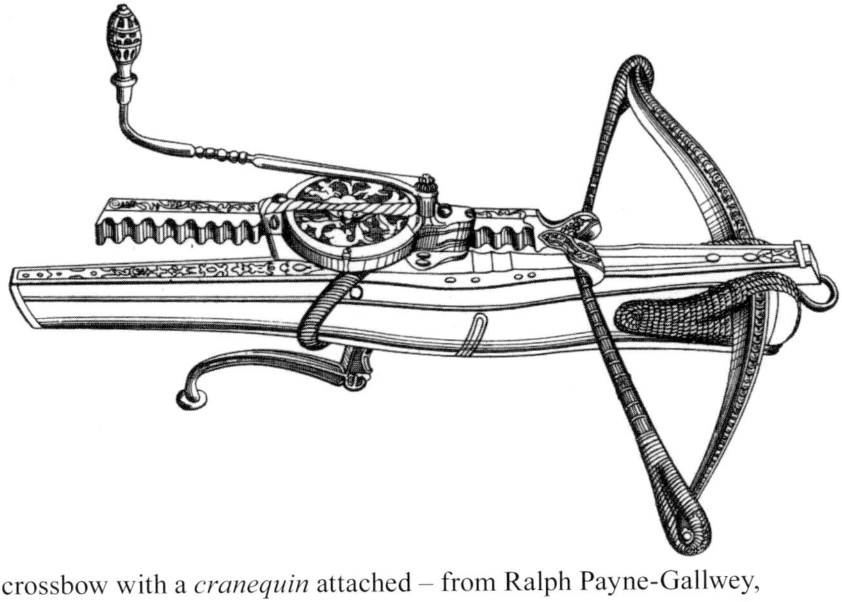

A crossbow with a *cranequin* attached – from Ralph Payne-Gallwey, *The Crossbow: Its Military and Sporting History, Construction and Use* (London, 1903).

manner similar to using a belt hook, see for example Plate 19. While not on the same level of burden as a breastplate or a shield it was still a significant piece of metal that the archer would have to hang from their body and therefore a potential source of fatigue in battle.[29]

There were a few important differences in how the *cranequin* and the windlass were used. The windlass spanned the crossbow in one motion which required constant cranking while the *cranequin* pulled the string back in distinct stages so if the archer paused the string would not change position. In between shots the windlass could be re-armed by pulling the ropes back to their starting position in a matter of seconds while the *cranequin* had to be cranked all the way back to its initial position, a slower and more laborious process. The *cranequin* was also more durable since it was made entirely of metal, and also more compact, but heavier as a result. Both devices were slow to operate which gave these powerful crossbows slow reload times. As with the windlass, an archer using a *cranequin* generally took between 30 and 60 seconds to load each shot.[30]

The power of spanning devices is measured by their advantage, which is the extent to which the force the user puts into the device is increased by the device's operation. Unfortunately, advantage is very rarely determined for medieval spanning devices. The only widely available calculations of advantage are by Jens Sensfelder and apply exclusively to *cranequins*. Sensfelder calculated the advantages of ten *cranequins* from the fifteenth and sixteenth centuries, three from the Royal Netherlands Armoury and seven from Grandson Castle. The smallest advantage provided by a *cranequin* Sensfelder measured was 1/57, meaning that the force pulling the string back was 57 times greater than the force required to rotate the handle of the *cranequin*. The greatest advantage was 1/342. Advantage is a useful metric for examining the power of crossbows since *cranequins* with very large advantages would only have been made if there were crossbows that needed that kind of power to span them.[31]

The screw

The screw was another method for spanning very powerful crossbows, but one that was rarer than either the windlass or the *cranequin*. It was aptly named, as it was essentially a large screw with a hook on the end and a handle to turn it. This handle would extend or retract the hook. To use it the archer would extend the clawed end until it captured the string, and then turn the handle the other way to pull the string was into place. It was sometimes built into the crossbow rather than being a separate removable piece of equipment, in which case the claw was rotated 180 degrees to clear it from path of the string before shooting.

The screw's origins are somewhat contested. Howard Blackmore suggested that this was an early spanning device, first being used with siege weapons before moving on to portable crossbows, and by implication predated the windlass. However, there is no solid evidence for this supposition, and it is probable that the screw and the windlass were contemporary solutions to the same problem. The *cranequin*'s design probably owes at least some elements to the screw, and it is possible that it largely replaced the screw. That said, like the other devices the screw lasted well into the early modern era, and all surviving examples of the screw are modern.[32]

The krihake

The *krihake*, also known as the Samson belt, was a hook with a pulley on a rope that ended in a metal loop on one end and hung from the archer's belt from the other end. To use the *krihake* the metal loop was attached to a metal lug or hook on the crossbow's stock and the *krihake*'s hook was attached to the bowstring. The crossbow was spanned by the archer placing their foot in the crossbow's stirrup and then straightening up to standing. The process of straightening up would pull the string back into place.

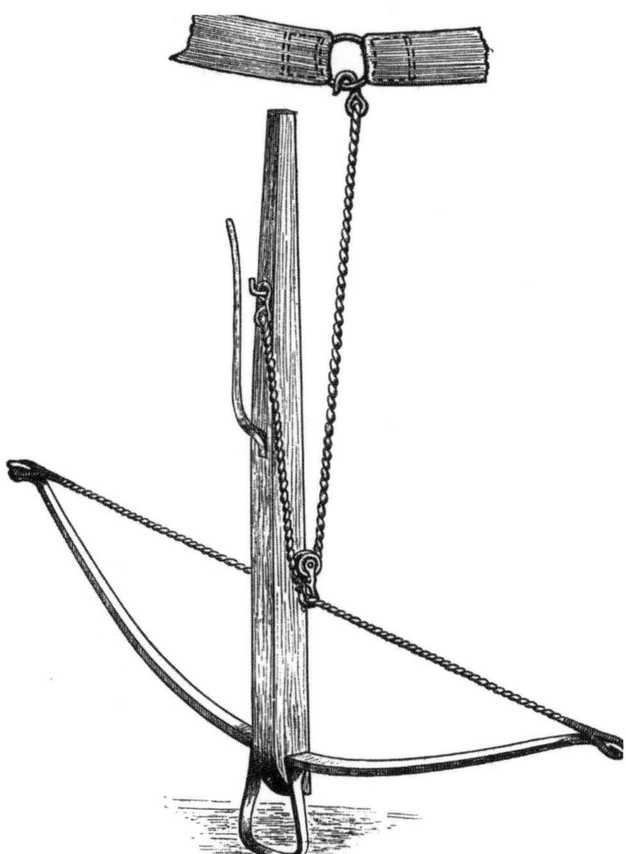

A crossbow with a *krihake* attached – from Ralph Payne-Gallwey, *The Crossbow: Its Military and Sporting History, Construction and Use* (London, 1903).

The *krihake* was similar to but also superior to the belt hook because the pulley and rope system used allowed for the archer to use their force more efficiently, which in turn allowed them to either span a heavier crossbow or to span a standard crossbow with less effort. Despite the relative simplicity of the *krihake*, especially compared to some of the devices discussed in this chapter, there is no evidence that it was an early invention. It was probably first introduced in the fourteenth century and then became common in the fifteenth and sixteenth centuries. There is ample artistic evidence of its use from the fifteenth century, perhaps most impressively in the Martyrdom of St Sebastian by Piero de Pollaiuolo (Plate 18).[33]

The goats-foot lever

The goats-foot lever was two sets of metal claws joined together by a hinge with a handle (Plate 4). It was usually used by hooking one set of claws over the string and the other over a metal lug in the stock. The handle was either pushed or pulled, depending on design, to lever the string into the lock. The goats-foot lever was in general use from the fourteenth century but how far back it dates before then is not clear. It continued to be used and new versions were even being developed well into the modern era.[34]

The goats-foot lever is the other medieval spanning device, besides the *cranequin*, to survive in complete examples in significant numbers. This is not particularly surprising given that both devices were made entirely out of metal, although in the case of the goats-foot lever it could be made out of wood or other materials as well. Surviving medieval goats-foot levers follow a broadly uniform design but with minor stylistic differences. Goats-foot levers were also very light, with an average weight of less than a kilogram and some as light as 500g. The main trade off with the goats-foot lever in comparison to other spanning devices was that the archer still had to do a large portion of the work to span the crossbow. For lighter crossbows this was not a problem, but for heavier crossbows this would have made the goats-foot lever impractical. In exchange for lower power the goats-foot lever offered speed and reliability.[35]

Josef Alm suggested that the goats-foot lever was invented to enable the reloading of a crossbow while mounted since it was light and did not require the use of the archer's legs. There is merit to this argument, but the goats-foot lever was used by more than just mounted crossbowmen and it was probably intended for use by a wide range of archers. Many surviving goats-foot levers have a hook on their handles that would have allowed them to hang from a saddle or a belt, further reinforcing the portability of the device. The goats-foot lever was also designed to fold up into a compact shape, making it significantly more portable than the either the *cranequin* or the windlass.[36]

Spanning a crossbow with a goats-foot lever – from Ralph Payne-Gallwey, *The Crossbow: Its Military and Sporting History, Construction and Use* (London, 1903).

The spanning stand

The spanning stand was a wooden stand that the archer mounted their crossbow vertically on with its bow facing towards the ground and string side facing the stand. The string was pulled back via a lever mounted on the top of the stand which was connected to a hook that was attached to the string. This system was very efficient at spanning crossbows and often used with very powerful ones, but its massive bulk made it difficult to move. This meant that the spanning stand was used primarily during sieges. Spanning stands were also very commonly used with great crossbows, a type of crossbow that was too large to be carried by an individual and usually used to defend castles and city keeps. That said, there are references to spanning stands that were taken on campaign and deployed in battle. The stand was also most effectively used by pairs of archers with one person reloading and the other shooting. The spanning stand has received far less attention than other more famous spanning devices like the *cranequin* or windlass, but it does not seem to have been less important in medieval warfare.

As has been the case with most of the spanning devices this chapter, the origins of the spanning stand are not well documented. There are plenty of references to them in the fourteenth century, including detailed records of their purchase by the popes in Avignon and several accounts describing the costs involved in hiring carpenters to build them. There are also several manuscript depictions of them in manuals of war, including Conrad Kyeser's (1366–c.1405) famous *Bellifortis*, which was written between 1402 and 1405.[37]

Crossbow Bolts

Any discussion of the crossbow must take some time to consider its primary ammunition: the bolt. Medieval bolts were generally shorter, thicker and often heavier than standard arrows. While most arrows, including medieval ones, had three sets of feather fletching, bolts were only sometimes fletched and when they were they usually only had two sets positioned on opposite sides from each other on the bolt. While they could be made with feathers they were also made from paper, leather, wood or even bronze. Bolts' shorter length and different weight distribution meant that they benefited less from the fletching than arrows did. Given the cost and time involved in properly fletching an arrow it should be no surprise that most military bolts seem to have not been fletched at all. When hunting or competing in target shooting, activities where accuracy was essential and a single bolt could be shot multiple times, fletching was more of a priority, but in a battle where hundreds of bolts were shot never to be seen again it was not.[38]

The most common type of military crossbow bolt head was the bodkin. The majority of these were roughly diamond shaped, which helped to direct the force of the bolt for maximum penetration while also being sturdy enough to survive impact. A slightly less common version was the needle bodkin. This was a narrow metal cylinder mounted on the end of the bolt, almost like a throwing dart. While the needle bodkin effectively funnelled all the power of the bolt into one point, their lack of structural support meant that on a glancing blow against an armoured target they would usually deform or break, making them most effective against unarmoured targets, although they could achieve significant penetration if they managed to hit just right.[39]

Crossbows could also be used to shoot incendiary bolts, although they were not usually used like they are in the movies. These bolts had long thin heads that ended with a small barbed tip. The long metal head would be wrapped in flammable material and set alight before being shot. The barbed tip enhanced the chances that it could latch on to something flammable when it landed. The long metal head ensured that neither the bolt nor the crossbow caught fire while they were being shot – always an important consideration. Incendiary bolts were rarely used in battle, their primary role being in siege warfare where they could be shot into a city to try and start damaging fires or as a means of burning up enemy siege machines.[40]

The widest range of bolt types can be found in the arsenal of the medieval hunter. The most popular of these was the barbed bolt, which was designed to penetrate deep into the flesh of large game and be impossible to dislodge without making an even larger wound, eventually killing the animal. Crescent-shaped bolts were used to shoot a limb off of smaller game, usually to leave them to bleed out so the hunter could collect their pelt later. There were even blunt crossbow bolts that could kill small birds and rodents by force of impact alone. These heads had the benefit of not damaging the pelts or meat of the animals they killed. Plate 5 shows a selection of the wide array of bolt heads that might be used by medieval hunters stalking their prey.[41]

The Physics of the Crossbow

One path to better understanding the use and performance of medieval weapons is through the principles of modern mechanics and ballistics. Medieval bows and crossbows are rarely, if ever, in a fit state to be shot by modern-day historians but with a good grasp of the mechanics of archery we can try to better understand how these weapons would have performed when they were new. Unfortunately for the contemporary historian, the physics of archery is so complex that a full understanding of it would probably require a second doctorate in physics.[42] Instead most historians must settle for a more general approach which can help guide the examination and understanding of surviving medieval crossbows.

The mechanics of the longbow are generally better understood than those of the crossbow. This is in a large part thanks to the popularity of traditional archery and bow making among modern hunters and archers. This has spawned a range of books on bow making which has in turn inspired a greater understanding of the mechanics of how a traditional wooden bow works. Recent work by scholars and crossbow enthusiasts, especially in the German-speaking world, is beginning to close this gap, but for the moment the bow is still the better-understood weapon. Due to the greater evidence for wooden bows much of the following section is derived from research on those weapons. The underlying mechanics of a crossbow function in much the same way as a traditional wooden bow – especially in the case of wooden crossbows. For composite and steel crossbows there are differences in how the weapons operate, but those differences only arise when you get very nuanced and specific in your analysis. Crossbows are at their simplest just short, powerful bows mounted on a stock and so many of the principles governing bows apply to crossbows as well.

A bow stores power when it is drawn due to the combination of compression of the belly of the bow and the stretching of the back. The greater the distance that the bow is drawn the greater the power that is stored.[43] When the string is released it transfers most of the energy to the arrow or bolt as the bow returns to its original resting position. Energy is also spent moving the bow's arms back to their starting position: the heavier the arms the more power required. Energy is also lost through other functions, such as string vibration or via the friction of the missile against the weapon's stock or handle.

The power of a wooden bow is primarily determined by its thickness, thickness in this case meaning the distance from the back of the bow to the belly. This is because when the bow is drawn more force is required to overcome the resistance of the thicker wood. A bow that is built thicker will greatly increase its power but since the wood can only take so much stress before splintering there is a limit to how much power can be gained this way. Increasing the width of a bow can help mitigate this problem as the force of the draw is spread across a greater area meaning no one section of bow takes too much strain.[44]

The different types of wood used in bow making have a significant impact on bow performance and design. Yew was the ideal wood for making both bows and wooden crossbows in medieval Europe in part because of the different properties of its sapwood and heartwood. The sapwood of yew is very elastic while the heartwood is very strong and rigid. The different properties of these parts of the tree function in tandem to allow for yew bows to be built much thicker than bows of other wood with minimal risk of breaking. It is no surprise that the majority of surviving medieval bows and crossbows are made of yew.[45]

Bows that are made of several different materials are known as composite bows. The mechanics of the composite bow are much better understood than those of the composite crossbow and there is an element of uncertainty about the extent to which the principles of one can be applied to the other. The 'standard' composite bow has a core of wood with a layer of horn along the belly. The bow is then wrapped in sinew with most of the layers along the back of the bow. When the sinew dries it shrinks and pulls the bow forward, giving it a natural resistance when drawn. The sinew is extremely elastic, which allows it to store a lot of energy while also preventing splintering or breaks when the bow is drawn. The horn along the belly is very resistant to compression which significantly increases the amount of energy required to draw the bow.[46]

Composite crossbows usually had more layers of sinew, especially along the back, and a very different core when compared to the composite bow. For an example of a composite crossbow core see Plate 1. When a composite crossbow was drawn the wood and horn compressed to generate significant power while the sinew stretched, generating more power by resisting being drawn while also holding the crossbow together. Composite crossbows came in a wide variety of styles and types, but the different performance of these designs is currently not very well understood. Hopefully further experimentation with replica weapons will help historians gain a better understanding of how the different types of composite crossbows functioned and how well they performed.[47]

The steel crossbow differed noticeably from previous examples since it was made of an inorganic material. It essentially functioned as a large metal spring that was put under two points of pressure at the ends of its arms. Egon Harmuth has argued that the best design was for thin steel bows of very high-quality steel, but safety concerns may have forced medieval smiths to build sturdier but bulkier weapons. Thicker steel limbs ensured against possible breaks but also had significant negative impacts on crossbow efficiency and power.[48]

There certainly seems to have been some fear around the possibility of steel crossbows breaking – something that could put the archer holding the weapon in significant danger in addition to rendering the weapon useless. The Teutonic Order was slow to adopt steel crossbows, preferring to stick with composite and wooden crossbows instead, possibly out of fear that the steel bows might break during their many winter campaigns. Arthur Credland has noted that the steel crossbow was popular in early modern Lapland which could suggest that the Teutonic Order's fears were unfounded but could also just reflect differences in weapon design making the later crossbows safer. In experiments conducted on an early modern steel bow, Jens Sensfelder found that reducing the temperature to minus 15 degrees Celsius increased the bow's draw weight by approximately 12 per cent. This increase could have caused damaging breaks, especially over repeated uses. In 1455 Gilles de Bouvier wrote that horn crossbows, meaning composite crossbows, did not break when frozen and instead only became more powerful, so it is possible that while Sensfelder is correct in noting the increase in power that could potentially have been seen as a benefit by some.[49]

Draw weight

The power of a bow is generally expressed in terms of its draw weight. Draw weight is the amount of force required to pull the string back to a specific distance. It is often represented in the form of pounds at a distance, e.g., 100lbs at 28in draw. Most hand bows are designed with an ideal draw distance in mind while crossbows have a fixed draw distance based on where the trigger lock is placed. The easiest way to determine draw weight is to connect a force gauge to a winch and draw the bow or crossbow back to a specific distance. However, this can rarely be done with surviving medieval weapons due to the risk of breaking a very valuable artefact. Even in cases when it has been done, like with a selection of longbows found on the sunken Tudor warship the *Mary Rose*, damage to the weapon in the intervening centuries can render these tests unreliable. As a result, draw weight for historic weapons is usually estimated with the help of computer modelling, the construction of replicas and/or supplementary evidence, like data on contemporary arrows or bolts.[50]

Draw weight measures the power stored in the bow, but that power does not fully transfer to the arrow when the string is released. Energy is spent returning the limbs of the bow back to their original positions, in the vibrations of the string after firing, and in the friction of the arrow against the bow or stock. The string also stops before it reaches the bow due to the weapon's natural curve, which means that the projectile stops being pushed by the string before all the power stored in the bow can be transferred to it. The measure of how much of the draw weight is transferred to the arrow is the efficiency. The only way to reliably calculate the efficiency of a bow is to measure the draw weight and then shoot several shots from the weapon, calculate the speed of the projectiles and then compare that speed to a theoretical speed if the weapon were 100 per cent efficient. This is generally not practical with historic weapons, so historians have explored other methods, such as calculating the efficiency of replica weapons with the help of computer models or estimating the efficiency based on the bow's design and dimensions.[51]

In broad terms, crossbows are small, heavy bows so they have huge amounts of power but also are often very inefficient. W.F. Paterson determined the efficiency of a crossbow he tested to be approximately 40 per cent, meaning less than half of the draw weight of the crossbow was transferred to the bolt. There is still plenty of work to be done in this area though, and so this figure should not be taken as a representative average for the crossbow as a whole. Instead, it is but one very intriguing data point.[52]

Another point to consider is that there is often a limit to how fast a bow or crossbow can launch a projectile. Increasing power does not necessarily guarantee an increase in projectile speed. A crossbow with a high draw weight will often shoot a bolt at the same approximate speed as a less powerful crossbow. This means that to gain greater power the best option is to increase the weight of the bolt since the momentum of the bolt is its mass times its velocity. If you cannot increase velocity, then you must increase the mass instead.[53]

The historian and chemist Alan Williams included a thorough discussion of the capacity of medieval and early modern weapons to penetrate armour in his 2003 masterpiece *The Knight and the Blast Furnace*. By careful study of the design of plate and other metal armours, as well as the physics and chemistry governing their sturdiness, Williams was able to establish how much power a weapon would have to deliver to penetrate a given piece of armour. His study has provided the best account of all the variables involved in the attack on armour, and how each of those variables impacted the performance of the weapons involved. Williams described how an indirect angle of attack required more energy to cause penetration and how layers under armour increased the total energy necessary to deliver a damaging blow to the wearer. While primarily concerned with metal armour, Williams also included a brief discussion of leather and cloth armours.

Overall, Williams' study suggested that, as the Middle Ages progressed, armour became increasingly resistant to penetration by bows and crossbows. While bows and crossbows were generally capable of breaking the rings on a suit of chain mail, they would have struggled against plate armour. While they could probably penetrate the thinner arm and leg armour if they landed a good blow, breastplates and helmets would have been all but impenetrable for the average bow or crossbow.[54]

In his book on great crossbows and springalds the French historian Jean Liebel included a fascinating chapter on the ballistics of both weapons, as well as useful data on the handheld crossbow. He used mathematical models to show the stages of how power is built when a crossbow is spanned, including the amount of draw weight that is stored at various points in the spanning process. He then examined the force in joules that springalds and crossbows of various sizes would transfer when shot and then what their likely force on impact would have been. His work captures how powerful these weapons were, especially the heavier siege weapons which were capable of truly outstanding impacts.

The chapter also included a detailed discussion of what the force of impact would be when shooting the weapons at a variety of angles. Liebel's figures, which mostly agree with Williams', give the power of a standard crossbow as 126 joules, while for the great crossbow he estimated either 331 joules or 627 joules depending on the size and type. Liebel's work is nearly unique in its discussion of these larger weapons that were beyond the capacity of a single archer to carry or use. If there is one problem with Liebel's work, it is that it would benefit from greater clarity on how he calculated his initial figures for the draw weights of the weapons. It is not a major flaw, and the work in this chapter is of huge assistance to any student of the crossbow, but it is an area that could have been improved.[55]

Draw distance

One of the enduring mysteries of European crossbow design is why their draw distances are so short. I will note at this stage that I will be using draw distance for this discussion, but it could equally be framed in terms of power stroke. The power stroke is the section of the crossbow where the string imparts its energy to the bolt while the draw distance is the full distance between the bow and the trigger lock. The power stroke is always shorter than the draw distance as the string inevitably stops before becoming flush with the bow due to the bow's curve when strung. There are important factors to consider when comparing the two in a fully technical analysis, but for this section draw distance will suffice.

A traditional yew longbow had a draw distance between 26in and 30in. In contrast, a sixteenth-century steel crossbow had an average draw distance of 8.5in. The average draw distance for fifteenth-century composite crossbows seems to have been 8.6in. The Berkhamsted Bow, one of only a few surviving medieval wooden crossbows, has been estimated to have had a draw distance between 8in and 10in. There is nothing in the principles of crossbow design that necessitates these short draw distances and increasing the draw distances could have been a simple way to generate more power. Chinese crossbows, for example, had much longer draw distances.[56]

Draw distance is closely related to the length of the bow and it could be that for medieval crossbows a shorter bow was considered preferable to a longer one. For a steel crossbow the weight of the bow could have been a concern – a crossbow with a large steel bow would have been front-heavy and difficult to shoot. However, the issue of weight does not explain why composite and wooden crossbows also have such short draw distances. The consistency of the draw distances across all three types of crossbow makes this explanation unlikely.[57]

Another possible explanation is the rolling nut trigger. Because of the size of the Z-trigger the lock mechanism was generally located in the centre of the stock rather than at the end like in Chinese crossbows. The positioning of the trigger lock set a limit on the weapon's draw distance, and if the trigger had to be in the middle of the stock for ease of shooting, then that would have shortened the draw distance. However, this raises more questions. The Chinese bronze trigger was a much older technology, and it seems likely that someone in Europe must have encountered it at some stage. Again, we are left with an unsatisfactory explanation. The rolling nut trigger may very well have caused the short draw distances, but then we need to explain why it was adopted in the first place.[58]

For the time being the reason behind the short draw distance remains a mystery but further research and examination of a wider range of the surviving medieval crossbows will hopefully provide new insight and maybe even an eventual answer.

Experimental Archaeology and Crossbows

One of the ways to try and measure the power and performance of medieval crossbows is through experimentation with replica weapons in a practice known as Experimental Archaeology. There are many challenges facing any historian or archaeologist attempting to better understand the past by experimenting in the present, but it can be a very valuable process. Even when experiments are not perfect, which is most of the time since perfection is a near-unattainable goal, they can still provide fascinating insights and highlight areas for further study.

In 1998, the Royal Armouries published the results of a series of tests of medieval ranged weapons led by curator Thom Richardson. These tests involved a wide range of weapons from the Middle Ages and antiquity, including a replica of an arquebus from the *Mary Rose*, a medieval handgun, three longbows, two crossbows, six types of slings, and two methods for throwing a spear. The bows and crossbows were shot a number of times using ammunition with different heads. The speed of each projectile was measured, and then an average calculated for each type of projectile.

The testing showed little difference in the speed of the projectiles shot from either the crossbows or the longbows. The first crossbow had a bow made from yew with a draw weight of 90lbs (41kg). The second was a steel crossbow with a draw weight of 440lbs (200kg). The longbows were all made of yew and drew 72lbs, 78lbs and 90lbs (33kg, 35kg and 41kg), at 28in (71cm). The longbows shot arrows at 93mph and 97mph, while the steel crossbow achieved 98mph and 100mph with two different bolts. The wooden crossbow was closer to the longbows with speeds of 81mph and 93mph for two different bolts. All the weapons were also shot at a 2mm thick sheet of steel, but none of them were able to penetrate it.[59]

The primary objective of these experiments was to show the sudden increase in the power of early gunpowder weaponry in comparison to more traditional projectiles. It easily met this goal as the gunpowder weapons fired their bullets at a speed that was an order of magnitude higher than either the bows or the crossbows. The draw weights for the bows and crossbows used were lower than what would probably have been normal in the Middle Ages so the numbers should be considered a minimum threshold for medieval crossbow power. Even factoring that in, though, the power of the early firearms was so much greater that it is clear that crossbows could not compete in terms of sheer power.[60]

Experimental archaeology draws on a different set of skills than most other areas of history or archaeology and as such can attract practitioners from different backgrounds. Nowhere is this more apparent than on YouTube where there are numerous creators who make videos about medieval technology and weaponry, including archery and longbow techniques. While the quality of these videos can vary significantly, they often represent a different perspective on the history of archery than you would find from a traditionally trained historian.

One practitioner who is worth a watch for any student of the crossbow is Tod of Tod's Workshop. Tod is a skilled maker of replica medieval weapons including swords, bows and, of course, crossbows. In addition to his business as a weaponsmith, Tod has a YouTube channel where he films himself experimenting with his creations. Tod has made videos comparing shooting speeds between longbows and windlass-powered crossbows, testing the maximum range of a medieval crossbow, measuring bolt speeds for a variety of crossbows, and many videos of him just out shooting a variety of replica crossbows. Tod also tries to include experts in several of his videos, most notably in his collaborations with Dr Tobias Capwell, an expert in medieval armour and curator at the Wallace Collection in London. While hardly the last word on crossbow testing, Tod's videos are a fascinating glimpse into the operation of the medieval crossbow and provide insight into how these weapons would probably have been used. It is rare to find anyone who has spent a considerable amount of time making or shooting medieval-style crossbows in the modern era, and for that alone what Tod has to say is probably of interest to anyone studying the medieval crossbow.

Chapter 2

Types of Crossbow

The English have the dubious claim to having lost more rulers to archery than any other kingdom. This book opened with a famous example: Richard I and his death at the hands of a crossbowman of inconclusive identity. Among these dubious ranks also lie King Edmund the Martyr (r.855–869), who was supposedly tied a tree by Vikings in 869 AD and shot full of arrows, as well as King Edmund II (r.Apr–Nov 1016), possibly, and William II (r.1087–1100), definitely, whose deaths will be discussed in later chapters.

Left out of the above list is the king whose death by arrow is probably the most famous of them all: Harold Godwinson (r.Jan–Oct 1066). As most schoolchildren know, Harold was killed by an arrow to the eye at the Battle of Hastings. His death was the battle's turning point and forever changed English history. The general assumption in this story is that Harold was shot by a traditional bow and arrow, and that assumption is not without foundation. As with the death of Richard I, there are numerous accounts of the Battle of Hastings and, when they even bother to provide any details about it, they do not all agree on the specifics of King Harold's death. This leaves open the possibility that if Harold was killed by a shot to the eye, it could have been a crossbow bolt rather than an arrow that dealt the killing blow. However, that is a big if, as this section will show.

William of Poitiers (c.1020–90) was a chaplain to William the Conqueror (r.1066–87) and, prior to taking the cloth, had served as a soldier under Duke William. While he was not a participant, William wrote one of the most thorough histories of the conquest of England. He completed this work sometime around 1071, and while he was not an eyewitness to Hastings, he would have had the opportunity to interview people who were. His account has generally been regarded as reliable, if very pro-Norman. While generally very detailed in describing the battle, William did not provide any description of the death of Harold, stating initially that he had died along with his brothers and the leading men of his kingdom. In the aftermath of the battle William provided the further detail that Harold's body was found with two of his brothers next to it and had to be identified by 'certain signs' instead of by his face. He also mentioned that the bodies had been stripped of all insignia.[1]

While William made no mention of the involvement of arrows in Harold's death, he did describe the use of crossbows by the Norman invaders. William mentioned that the Norman vanguard consisted of infantry armed with bows and crossbows, and then later in his narrative he wrote: 'The Normans shot arrows, wounded and transfixed men; the dead as they fell, moved more than the living.'[2] Clearly archery played an important role in the battle and nothing in William's narrative directly contradicts the arrow story but beyond saying that Harold's dead face was unrecognizable he gave no evidence that could support it either.

A similar account that provides slightly more insight into Harold's death can be found in the poem *Carmen de Hastingae Proelio*. The text of this poem does not identify any author, but it was initially attributed to Guy, Bishop of Amiens (d.1075) because the twelfth-century author Orderic Vitalis (1075–c.1142) mentioned that Bishop Guy had written a poem about Hastings and no other evidence of that work survives. If Bishop Guy were the author, it could place the *Carmen* as early as the first half of 1068, making it the earliest account of the battle. Given its similarity to William's text, it has been suggested that this may have been a source he drew on when writing his version. However, the identification of Guy as the author was challenged by R.H.C. Davis who argued that the style of the work, as well as some of its details, meant that it likely dated to 1100 at the earliest, meaning it was inspired by William of Poitiers and not the other way around.

The case is not settled, but the difficulty in dating of the poem as well as its tendency for fantastical literary flourishes means it should be read with a healthy dose of scepticism. Like William of Poitiers, the author described the deployment of crossbowmen in the Norman vanguard and even says that they were deliberately aiming for the faces of the English soldiers. However, for Harold's death it describes an attack by four Norman knights, including Duke William, in the midst of a vicious melee, who hack the king to pieces. This is the most dramatic of the early versions of Harold's death, and while the text does not credit William with the killing blow it does give the duke and future king a central role in the regicide. Jim Bradbury views this as further evidence to discount the version of events in the *Carmen* since no other authors mention William playing a role in Harold's death and he cannot imagine why they would leave out so important a detail if it were true. Suffice to say, the *Carmen* contradicts the story that Harold was shot in the eye, but there are reasons to doubt the validity of its version of events.[3]

The Norman monk William of Jumièges (c.1000–after 1070), like William of Poitiers, did not participate in the conquest of England but wrote an account of the campaign in or around the year 1070. His account of the campaign and battle are both very short, doing in a few lines what William of Poitiers took pages to describe. No mention is made of archers, and of Harold's death he wrote: 'Harold himself, fighting amid the front rank of his army, fell covered with deadly wounds.'[4] Again, nothing to rule out death by arrow but nothing to suggest it either.

To get an English perspective on the battle one must turn to the *Anglo-Saxon Chronicle*, which is really several similar works compiled in different parts of England from a shared tradition of annals. Three versions of the *Anglo-Saxon Chronicle* describe the events of 1066, but only two of these mention Hastings and neither in detail. All three give more time to the Battle of Stamford Bridge, fought earlier in 1066, at which Harold defeated invaders led by the Norwegian King Harald Hardrada. The two that do include Hastings mention the death of Harold and his brothers at the battle, but do not provide any details of the circumstances.[5]

To find evidence of the arrow shot to Harold's eye, historians must look to later sources, including the most famous of them all: the Bayeux Tapestry. Technically an embroidered cloth rather than a woven tapestry, it was probably commissioned by William the Conqueror's half-brother Odo, Bishop of Bayeux (d.1097). It seems to have been made in England and completed c.1077. While some historians have argued that it was made later, it is generally agreed that it was sewn within a generation of the battle. In this version of Harold's death, it shows an individual with an arrow in their eye and text above them that says, 'Harold Rex Interfectus Est' which means 'Here King Harold is Killed'. There has been some debate among scholars about whether this text should be matched to the individual with the arrow in the eye or to the one next to him who is being hacked down by a mounted soldier wielding a sword.[6]

David Bernstein has suggested that these scenes should be read in sequence and that both figures are Harold. The first image is him being shot in the eye and the second is a wounded Harold being killed by a Norman knight. Bernstein examined the tapestry in detail and argued that there are signs on the original to indicate that the second figure being hit with a sword also once had an arrow in his eye, but the thread has been lost to time.[7]

However, the Bayeux Tapestry is nearly a millennium old, and in that time it has suffered wear and tear and, importantly, it has also been restored and repaired. Several eighteenth-century engravings of the tapestry do not show an arrow in the all-important death scene, and it is possible that the current arrow was inserted sometime in the nineteenth century as part of a restoration of the tapestry. That restoration may have correctly replaced a missing arrow clearly visible from holes in the fabric, but it also may have replaced a large projectile like a spear or even relocated the arrow from elsewhere on the figure's body to its eye. Similarly, the holes found by David Bernstein could have appeared in the tapestry sometime after the eleventh century.[8] The Bayeux Tapestry is a difficult document to interpret and historians must remain open to the idea that either of the two figures could have been intended as King Harold by the original creators.

The most solid evidence for the arrow in Harold's eye comes from accounts written in the generation after the battle. It would take too long to recount every example from this period, so instead two particularly famous authors will be considered: William of Malmesbury (c.1096–1143) and Henry of Huntingdon (c.1088–1157).

William of Malmesbury was born in Wiltshire to a Norman father and an English mother. He was one of the most influential medieval English historians and wrote multiple books during his lifetime. In his *Gesta Regum Anglorum*, or *Deeds of the English Kings*, he described the lives of English kings between 449 AD and 1127 AD. In his account of Hastings, he claimed Harold was killed by an arrow to the brain with the added detail that a Norman knight later slashed the dead king's thigh with a sword, bringing much shame on himself for desecrating Harold's body and being dismissed from King William's service as punishment. The way William described the death-dealing arrow included a reference to Virgil's *Aeneid*, and it could be argued that his account owes more to classical poetry than reality. It is also possible that William was correct about Harold being shot but changed a crossbow bolt into an arrow to better fit his classical allusion.[9]

Henry of Huntingdon was an English monk born in Lincoln after the conquest. His *Historia Anglorum*, a history of England from its beginning until 1154, is one of the most important and influential accounts of the early period of Norman-ruled England. While Henry was born after the events of Hastings, he drew on earlier accounts in constructing his narrative. In describing Harold's death, Henry wrote: 'Meanwhile, a shower of arrows fell round King Harold, and he himself was pierced in the eye. A crowd of horsemen now burst in, and the king, already wounded, was slain.' Both Henry and William's versions, written the generation after the battle, include the arrow to the eye and a knight slashing the king.[10]

While only two accounts made a specific reference to crossbowmen participating in the Battle of Hastings, the *Domesday Book* provides some further insight into crossbowmen in the early days of Norman England. It was compiled c.1086 by the new Norman monarchs to record exactly who possessed what land and what they owed in service for those lands. Its level of detail and scope makes it one of the most valuable sources in all medieval history.

There are several mentions of archers in the *Domesday Book*. In translation these archers are often given the same title but in the original Latin something else emerges. Take for example Nicholas the Bowman, a reasonably prosperous landholder in Warwickshire. In the original Latin he is called Nicholas *Balistarius*, which means Nicholas the Crossbowman.[11] There are at least two other archers that examination of the original text reveals were crossbowmen. While there is no direct evidence linking these individuals to the Battle of Hastings – the *Domesday Book* was compiled years after the battle – it is not a great leap of imagination to suppose that some of these crossbowmen received their lands as a reward for service during King William's invasion. These men could have been commanders of the crossbowmen mentioned in William of Poiters' account of the battle.[12]

It is impossible to be certain about any of this. Was Harold shot in the eye? If so, was it with a bow or a crossbow? The strongest evidence for the story of the eye shot is the Bayeux Tapestry, but that version clearly shows that it was an arrow, not a crossbow bolt. However, the Bayeux Tapestry does not include any crossbowmen at all, and we know from William of Poitiers that crossbowmen participated in the battle. It is possible that in dropping the detail of the crossbowmen the Bayeux Tapestry could have dropped a detail like the presence of a crossbow bolt in the face of a certain English monarch. We also know that the Bayeux Tapestry has been modified several times, and the only real distinction between an arrow and a bolt is the length. Only one of the contemporary accounts provides a version of Harold's death that precludes him having been shot in the eye, and the dating and accuracy of that account is disputed. Historians have debated whether the arrow shot happened for decades and there is little indication that they will reach a universally-accepted conclusion on the matter anytime soon. Realistically, we will probably never know for certain how Harold died: it will remain a much-debated mystery for the rest of human history.

Historians try to be as certain as possible about the past but medieval accounts all too often leave out details that were not thought to be important at the time. This applies just as much to the history of the crossbow as it does to the story of how a famous monarch died. Crossbows came in a range of types and were called by an even greater number of names, and untangling these types and names is a puzzle that has thwarted many historians.

Medieval Crossbow Terminology

The previous chapter examined the classification and design of crossbows using the framework that most modern historians employ. This system emphasizes evidence from surviving medieval crossbows and relies upon a modern perspective on how they should be classified. However, a different perspective can be found by studying medieval records and the writings of medieval authors. One of the challenges with this approach is that the terminology used by medieval authors to describe crossbows was often vague and rarely included any physical description. The meaning of terms also changed over time, for example the use of the word *ballistarius* for crossbowmen, common at the time of the *Domesday Book* discussed above, had largely fallen out of favour by the reign of Henry II (r.1154–89) some 80 years later. In its stead English records had adopted the French-derived term *arbalasterii*. This presents a challenge to any historian trying to study the history of the development of the crossbow via medieval records. With the meaning of terms changing over time and between geographic regions, piecing together a coherent vocabulary for the crossbow is an enormous challenge, and that is before any attempt to study what is being said with that vocabulary.[13]

One of the most widely-discussed cases of confusing medieval crossbow terminology comes from the thirteenth and fourteenth centuries. During this time numerous records from across Europe referred to certain crossbows as being either 'one-foot' or 'two-foot'. There were several distinct Latin and vernacular forms that were used, medieval language was not standardized and differences were especially pronounced in different regions of Europe, but the general gist was always the same: some crossbows were described as one-foot and others as two-foot. What they meant by those terms is not clear but there are some points that historians have generally agreed upon, the first being that one-foot crossbows were more common than two-foot crossbows. The second, which has slightly less evidence to support it, is that two-foot crossbows appear to have been more expensive than one-foot crossbows. In his study of crossbows in English financial records David Bachrach also noted that, in England at least, two-foot crossbows were more likely to have composite bows than one-foot crossbows were, but neither crossbow had a monopoly on bow type. This fact, combined with the greater cost, has led historians to speculate that two-foot crossbows were generally more powerful than their one-foot counterparts.[14]

There was a third type of crossbow that appeared alongside one-foot and two-foot crossbows in records of this time: the crossbow *ad turnum*. This was by far the rarest of the three weapons and it has generally been interpreted as a crossbow that was spanned with a windlass – a technology that was probably relatively new during the thirteenth century. The core of the historical debate around the meaning of *ad turnum* has been more to do with whether the winching system was physically built into the crossbow or was a detachable piece of equipment – by the later Middle Ages windlasses were separate pieces of equipment that latched on to the crossbow's stock, but it is not clear if they were always this way. Sir Ralph Payne-Gallwey believed the crossbow *ad turnum* referred to a spanning device akin to the *krihake* rather than the windlass, but this position has generally been rejected as there is no evidence for the *krihake* before the fourteenth century while the windlass was described by al-Tarsusi in the late twelfth century.[15]

While there is some general agreement among scholars about the meaning of *ad turnum*, the same cannot be said for one-foot and two-foot crossbows. The terms were usually used in financial accounts which do not provide context to clarify the meaning of the terms and no complete crossbow survives from this time, so it is not possible to pair either term with an archaeological example. Instead, historians have been forced to speculate. Guy Wilson, former head curator of the Royal Armouries in the United Kingdom, wrote an excellent article on the subject some years ago in which he outlined the three main theories of the meaning of these terms, as well as the problems underlying each one.[16]

The first theory is that the names refer to how the archer used their feet to span the crossbow. Supporters of this theory generally agree that for the one-foot crossbow the archer placed one foot in the stirrup and either spanned the crossbow by hand or using a belt hook. Where things get a little messy is with the definition of the term two-foot crossbow. Some proponents of this theory suggest that for the two-foot crossbows the archer placed both of their feet on the weapon's bow, one each either side of the stock, and pulled the string back in to place. This is a spanning technique we know was used. However, it is also generally thought to be an early technique, one that predated the widespread adoption of the stirrup since standing on the bow was both awkward and ran the risk of damaging it.

A similar interpretation is that the feet referred to the size of the stirrup, so some stirrups were for one foot and some for two. This theory has the problem that no stirrup wide enough for two feet has ever been found, as well as the fact that balancing with two feet next to each other in a stirrup would have been awkward and likely offer no advantage in terms of spanning power. This theory has one further challenge to it: a medieval document containing an order for crossbow equipment which included a set of three-foot crossbows – no one has yet been brave enough to argue for the existence of three-legged crossbowmen.[17]

The second theory is that the feet referred not to a quality of the archer, but rather to the length of the bolts. There are several references in English and French financial accounts mentioning the ordering of bolts for one-foot, two-foot or *ad turnum* crossbows specifically, which suggests that these weapons could have shot mutually exclusive sizes of bolt. This theory argues that the difference, at least for one-foot and two-foot crossbows, was in terms of the length of the bolts. However, not every record for the purchase of crossbow bolts distinguishes between bolts for one-foot or two-foot crossbows; some even mention the order of those types of crossbows but just mention the order of bolts for both weapons generally. There is a possible solution to this problem in an article written by Thom Richardson discussing the fourteenth-century springald, a siege weapon closer to the ancient Roman *ballista* rather than a crossbow. Richardson found that these springalds could shoot two different lengths of bolt. It is entirely plausible that the same was true of crossbows: they could have shot a specialist bolts as well as a more generic munition-quality bolt.

The other problem for this theory, though, is the complete lack of surviving medieval bolts that are 2ft long. In fact, there are very few bolts that are even 1ft long. Most surviving medieval bolts seem to have been a little shorter than a foot, with the added qualifier that the medieval foot and the modern foot are not exactly the same. There are plenty of surviving bolts but only some of them have been analysed, plus dating them to a specific century is a fraught process, so it is possible that there exists a collection of 2ft-long bolts from the thirteenth century sitting in a museum basement somewhere, but that does not seem very likely.[18]

The final problem for this theory is that there are perfectly valid alternative explanations for why one-foot and two-foot crossbows could have used different ammunition other than length. If we return to the earlier point that it is generally assumed that two-foot crossbows were more powerful, then they would probably have shot a heavier bolt as one of the greatest benefits to a heavier crossbow was its capacity to shoot a heavier bolt without losing any projectile speed.[19]

The third theory Wilson discussed is that the feet referred to some aspect of the crossbow's length. The most common version of this is that it meant the length of the bow. The lack of surviving crossbows from this period makes confirming this difficult. The closest example that survives is the Berkhamsted Bow, which is approximately 4ft long. It could be argued that the two-foot meant the length either side of the stock, in which case the Berkhamsted Bow would fit perfectly. The problem with this theory is that composite crossbows were generally much shorter than wooden bows, but we know that one-foot and two-foot crossbows came in both wooden and composite types and that it was the two-foot crossbows that were more likely to be composite. An alternative explanation is that the foot referred to the length of the stock, but one foot would be very short for a stock, and it is hard to see how the length of the stock would have made a big-enough difference to the crossbows performance to have been worth specifying so often in financial records.[20]

The mystery of what medieval writers meant by one-foot and two-foot crossbows still lacks a satisfactory explanation. The above should hopefully illustrate the complexity involved in interpreting medieval terminology. It is worth stressing that the terms one-foot, two-foot and *ad turnum* were only in use for approximately two centuries. They do not represent the full scope of medieval terms for crossbows nor were they even the exclusive terms in use at the time.

One final explanation for the meaning of one-foot and two-foot is probably the simplest but also the most frustrating; they probably did not mean one thing. Terminology can be frustratingly inexact, even in the modern era, and it often changes across time and space, so it is very possible that medieval authors meant several different things by their use of these terms. It is even possible that some of the above theories may have been right at some point.

The range of terms used to identify or describe crossbows used in the Middle Ages was vast and complex. It would be impossible to fully describe it here, but it is worth making some effort to represent a sample of what a historian who digs deep into this subject might find. Take for example the records of the Teutonic Knights in the Later Middle Ages. These usually used the term *armbrost* to refer to crossbows, which is still the word in modern German. However, in addition to *armbrost* you can find references to crossbows based on how they were spanned, such as the *stegeryffarmbrost* or the *wyndearmbrost*. These were, respectively, a stirrup crossbow and a crossbow spanned by a winder, presumably a windlass or *cranequin*. Terms could also denote power or size, such as the *ruckarmbrost* or the *gross windearmbrost*, the latter of which seems to have been a wall-mounted great crossbow. The *fuerarmbrost* seems to have shot incendiary bolts. There were some terms where the meaning is less clear, however, such as the *shutczenarmbrost*, *knottelarmbrost* and *zcylarmbrost*, in which cases, much like with the one-foot and two-foot crossbows, historians have to guess their meaning.[21]

Some German terms for steel crossbows with origins in the Later Middle Ages are still used by historians today. There are three main types of crossbows in this system, the *Rüstung*, sometimes also known as the *ganze Rüstung*, the *halbe Rüstung* and the much rarer *Viertalrüstung*, otherwise known as full-size, half-size and quarter-size weapons. Dirk Breiding has suggested these terms were derived from classifications for distances of target-shooting competitions. Despite what their names may suggest, the *halbe Rüstung* seems to have been the most popular size by far, with the *ganze Rüstung* having largely fallen out of favour after the Middle Ages. A fourth type of crossbow called the *Schnepper* was also popular, the distinction here seeming to be a different kind of trigger system. The *Schnepper* used a snap lock trigger as opposed to the rolling nut, and was also a lighter weapon that could be built to shoot bullets or bolts. Sometimes the bullet-shooting variety was known as a *Kugelschnepper*. These are by no means every German term used to classify crossbows, but rather they are a sampling of the variety a historian might have to confront in their study of the weapon in just one language.[22]

It is important to keep in mind when studying the medieval crossbow that modern classifications are the product of modern minds. Medieval minds had a different perspective. This can be seen in the different terminology used for these weapons, as well as the many kinds of crossbows that were produced. Much of this book will focus on the 'standard' medieval military and hunting crossbow, the portable type with a simple rolling nut trigger that shot bolts, but before we move on to discuss that weapon in more detail it is worth considering some of the more interesting and unusual types of crossbows that emerged during the Middle Ages.

Interesting and Unusual Crossbows

Bullet crossbows

The exact origins of the bullet crossbow have been hotly debated among crossbow historians. While most would date them to sometime during the late Middle Ages, whether they should be placed at the start or the end of that period has been contested. Josef Alm, usually the best source for crossbow history, believed that bullet crossbows originated in China and were probably brought back to

Europe by Portuguese merchants in the sixteenth century. However, there are clear textual references to the bullet-shooting crossbow in earlier centuries. In 1327 the people of London were reportedly plagued by hooligans shooting down the streets with their bullet crossbows. The weapon was not limited to hooligans, however. Its favour with royalty can be seen by the purchases of bullet crossbows made by King Louis XI (r.1461– 83) in 1479. It was also referenced in some fifteenth-century copies of Gaston de Foix's (1331–91) book on hunting, the *Livre de Chasse*.[23]

The bullet crossbow, sometimes known as a stone crossbow, was a crossbow that shot bullets, small round pellets or stones rather than bolts. The oldest examples were generally small with steel bows that were easily spanned by hand. The string was fitted with a pouch at the centre which held the bullet, and the nut was modified to grip a loop at the back of the pouch. The earliest surviving examples all seem to be Italian, and their stocks have a distinctive arced shape to prevent friction when the string is released. For example, see Plate 6 for an Italian-style bullet crossbow that was made in France, showing that the term reflected the origin of the weapon's style rather than that of the weapon itself.

Bullet crossbows were used for sport, either for target shooting or hunting birds and other small game. They were popular with women and children, but they were by no means restricted to those two groups. They were also quite faddish, surging in popularity in one region of Europe before fading away, only to reappear somewhere else. They were very common in Italy during the late Middle Ages and at the start of the Early Modern Era but almost completely disappeared from the seventeenth century. In Germany they were in use during the early modern era but fell out of favour around the same time as in Italy. They most recently had a significant major surge in popularity in England at the start of the nineteenth century.[24]

German bullet crossbows

While the Italian bullet crossbows had their very distinct arced stock, in Germany crossbow makers built a very different style. German bullet crossbows were made with a steel bow and metal tiller. The only non-metallic piece was a short wooden butt on the end of the tiller. German bullet crossbows also often had a claw spanning mechanism that could be operated by a lever built into the stock. These crossbows were magnificent examples of early modern engineering but that came at the cost of being much heavier than the Italian bullet crossbow. Like the Italian examples, they were exclusively intended for hunting and target shooting. While referred to as the German style of bullet crossbow, that just points to its region of invention. Both the German and Italian-style bullet crossbows were made in various regions across Europe during the heights of their popularity. For an example of a German bullet crossbow see Plate 7.[25]

Balestrini

While German crossbow makers may have shown off their engineering prowess with their bullet crossbows, Italian smiths took a different approach to making crossbows entirely out of metal. The *balestrino* was a metal crossbow which was small enough to be held in the palm of your hand. Despite their small size some had draw-weights of up to several hundred pounds, meaning the darts they shot could pack quite a punch. They were spanned by a clever built-in screw mechanism which was powered by twisting a section attached to the end of the stock. No *balestrini* survive from before 1500 and it is likely they were invented sometime during the early sixteenth century. For an example of an early modern *balestrino* see Plate 8.

Balestrini have sometimes been referred to as 'assassin's crossbows' but there is no evidence that they were ever used for assassination, nor even that they were ever intended for that purpose. The more likely explanation is that *balestrini* were intended for personal amusement and to show off the engineering talents of Italian artisans. There were clearly some concerns over their danger though, as Venice banned the possession of *balestrini* in 1545. In the seventeenth century they threatened the death penalty for anyone who was found with one. Clearly the ban did not work since there are multiple surviving *balestrini* from the sixteenth and seventeenth centuries. They largely disappeared after the seventeenth century, so either bans like Venice's began to take effect or the enthusiasm for the weapon waned.[26]

Great crossbows

Great crossbows were crossbows that were too large to be carried by a single archer but generally smaller than full-sized siege weapons like the springald. They were a major component of medieval warfare but due to the limitations of surviving evidence they are easily overlooked. Great crossbows only rarely featured in medieval art, with only a few notable appearances in famous books of war like Conrad Kyeser's *Bellifortis* and the works of Walter de Milemete (fl.1326), and the archaeological record is even worse. Only a handful of possible great crossbows have survived to the modern era, including one in Ingolstadt in Bavaria, one in Quedlinburg in Saxony-Anhalt, and one in Wrocław in Poland.[27]

Textual references to great crossbows are also fraught as 'great crossbow' was not a universal term in the Middle Ages. Instead, a variety of terms could be used to mean a great crossbow and historians must carefully read for context to understand the terms used. In some cases, these terms could be clear, for example one poet simply called them 'huge crossbows', but other times they can be difficult to decipher. Thankfully, the French historian Jean Liebel has done most of the hard work. Liebel also believed that 'two-foot' crossbows, discussed above, probably referred to a type of great crossbow.[28]

At its simplest the great crossbow was a large, powerful crossbow. Liebel suggests that their bows were between 1.6m and 2m (5ft to 6.5ft) long. Their bows

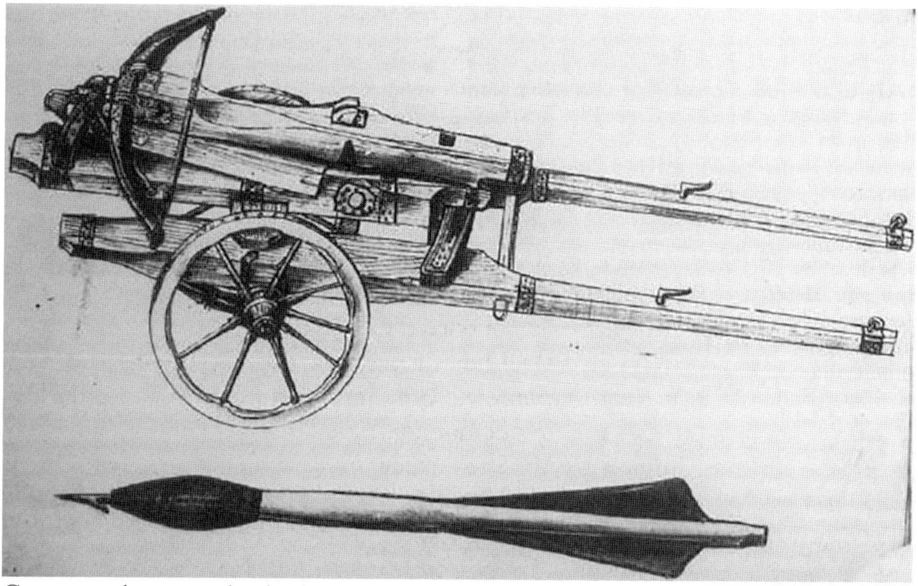

Great crossbow on wheels along with an incendiary bolt, MS lat. 599 fol. 34v, Bayerische Staatsbliblothek, Munich. (Public Domain Mark, Wikimedia Commons)

could be wood, composite or steel. They could not be spanned by hand alone and were most commonly spanned by a windlass, but a spanning stand, *cranequin* or screw could have been used instead. They were mounted on a table or cart since they were too heavy and cumbersome for an archer to shoot. Sometimes they were attached to their stand, but other times they were removable – this was necessary if they were used with a spanning stand, for example. Great crossbows were a class of crossbow and there was significant variety in the size and power of these weapons, just as there was between the types of hand crossbows.[29]

The fact that they required advanced spanning technology to be effective means that great crossbows most likely only became viable in the late twelfth or early thirteenth centuries. Their popularity seems to have fallen off in the fifteenth century, but they remained in use until the sixteenth century. Their size meant they were best suited to siege warfare and much of the available evidence for them is found in the accounts and finances of cities and fortresses. They were not entirely absent from battlefields, but they were overwhelming used from a fixed shooting position in towers, on walls and over gates where they were used to take shots at enemy siege weaponry and the crews operating them.[30]

Trap crossbows

The trap crossbow was a very specific type of hunting crossbow that was popular in northern Europe. Its origins lie in the Middle Ages, but it continued to be widely used well into the modern era, particularly between the seventeenth and nineteenth centuries. It was primarily used to hunt small game, including foxes, lynxes, badgers or martens. These were animals hunted primarily for their pelts rather than for food or sport which gives some indication of the type of hunter that was most likely to use a trap crossbow.[31]

The way a trap crossbow worked was relatively simple. The weapon was spanned, and a bolt put in place. They would often use a bolt with a crescent head, sometimes called a 'fox-knife' head which was designed to remove a limb from the creature in its sights, causing a slow death by blood loss and preventing the prey from running away. Its trigger was then connected via a cord to a trap that would cause the weapon to shoot once the target stepped on it.

Trap crossbows always had steel bows since they were left out in the open and rain would cause greater damage to other types of bow. Wood could lose its power if it became too wet and composite crossbows were often made with water-soluble glue – something that was usually not a problem for most archers but could cause problems for a weapon left out in the woods for an extended period of time. There was of course a risk of rust in steel bows, but that was more of a long-term problem. In some cases, the string of a trap crossbow was made of an iron chain, further reducing the risk of the weapon degrading while waiting for its prey. These metal strings were usually made of three pieces, two long pieces on each end and one smaller piece in the centre. Crossbows with steel 'strings' usually also had steel nuts to hold them. Trap crossbow strings could also be made of more traditional materials like hemp, which if covered in a thick layer of animal fat would be safe from damage due to moisture.

Trap crossbow stocks were always very short. The weapons were usually spanned using a *cranequin* and the stock almost always ended right after the *cranequin* lug. In some cases, these were purpose-built weapons, but it was also common practice to cut down older hand crossbows and adapt them for use as trap crossbows. In this way, late medieval crossbows were sometimes used as early modern trap crossbows. Trap crossbows were nearly always purpose built from the seventeenth century on while in earlier centuries it is more common to a mix of modified hand crossbows and purpose-built ones.[32]

Gun crossbows

In the author's opinion the most interesting development in crossbow history was the short-lived gun crossbow. These weapons combined a steel crossbow with the then very new wheel-lock gun. These had to be wheel-locks since that trigger mechanism did not require any form of open flame to ignite the gunpowder. Nobody, so far as we know, tried to attach a gun with a burning match to a crossbow. Gun crossbows first appeared sometime after 1500, the wheel-lock itself being an early sixteenth century invention. They seem to have been very rare, which is probably explained by how expensive they would have been to make. In total a little over a dozen gun crossbows have survived to the modern era.

Interestingly, three gun crossbows now in the Doge's Palace in Venice are probably the oldest surviving examples of wheel-lock gun mechanisms, which suggests that gun crossbows were among the first purposes this new technology was put to. Gun crossbows varied significantly in size; most were the size of a standard steel crossbow, but some were closer to being the size of pistols, a contemporary technology also made possible by the invention of the wheel-lock.[33]

It is not clear what the purpose of the gun crossbow was. They were far too expensive and rare to have been intended for mass use in contemporary armies. Some surviving crossbows shot bolts while others were built to shoot bullets (Plate 9), which suggests that they were for hunting since bullet crossbows were exclusively used for sport. One possible explanation is that they provided for redundant systems. Having a crossbow could have provided a hunter with a chance of still making their shot if the wheel-lock misfired. This would have been quite a niche situation though, as it would require both the weapon to misfire and for the target to not be startled by the sound of the gun misfiring. Would it not be simpler for an elite hunter to simply have a page carrying a backup crossbow?

Another suggestion, put forward by Dirk Breiding, was that the early examples in Venice may have been intended for use by bodyguards. This makes some sense; the redundant mechanism would be more valuable in a life-or-death situation. The wheel-lock gun would have provided more killing power than the crossbow, but with the risk of misfiring, so having the crossbow as well could ensure that at least one projectile would hit the target if both were shot simultaneously. Even if this was the original intention behind their invention, later weapons, such as the bullet crossbow examples, must have deviated from this purpose.[34]

A simpler explanation for why these weapons exist can be found in their very nature. They are masterpieces of engineering and design, and that alone may have been their appeal. A gun crossbow would allow artisans to show off their mastery and patrons their wealth.[35]

It is also important to view these weapons in their context. The sixteenth and seventeenth centuries saw a fad in adding guns to all sorts of traditional weapons. Gun swords, gun axes, gun shields, gun swords hidden in canes, hidden guns in books, and even gun cutlery were all made in this era. None of these weapons became a common feature of either war or sport in subsequent centuries, but the enthusiasm for wheel-lock, and later flintlock, gun mechanisms clearly extended beyond the realm of elite crossbow makers. Arguably the gun crossbow was one of the most practical examples of this trend. The invention of the wheel-lock opened a huge range of possibilities for the design of guns and all of these weapons could have been experimentations with this new technology first and practical weapons second. These may not even have been the first combination weapons. The Muslim author al-Tarsusi wrote a description of a combination spear-crossbow in the twelfth century, and while sadly no examples of this fascinating weapon have survived his description shows the enduring fascination with combination weapons.[36]

Chapter 3

The Crossbow in Art

The crossbow was a versatile weapon with many potential applications but one of the most bizarre is shown in a series of images in a late thirteenth-century Iberian manuscript of a work known as the *Cantigas*. Six panels are laid out over a single page of vellum much in the manner of a modern comic strip. They show the siege of a city by Muslim attackers, one of whom shoots his crossbow and hits a defender just below his right ear. The second panel shows a surgeon's attempts to remove the bolt with a pair of forceps, his failure resulting in the extreme steps taken in the third panel. The surgeons rigged up a large crossbow which they connected to the bolt still protruding from the poor man's head – presumably, they meant to shoot the crossbow in an attempt to pull the bolt free. That they are ultimately unsuccessful is shown over the next three panels as the man, bolt still firmly in place, prays for intercession and is ultimately spared his suffering by divine intervention.[1]

This practice may have been more common than one would think. In his early fourteenth-century book on surgery, Henri de Mondeville (c.1260–1320) claimed that the crossbow was a useful implement for arrow extraction on occasion and that he had only seen it fail once. While Henri's text provides slightly more detail, it is the images of the *Cantigas* that really drive home the dire circumstances that would entail this procedure and how horrible it likely was to experience.[2]

The Art of the Crossbow

One of the greatest challenges facing historians of the medieval crossbow is the limitations of the surviving evidence. While there is ample textual evidence for the use of the crossbow, which will be discussed in future chapters, texts that include detailed descriptions of the weapon are rare.[3] Crossbows survive in greater numbers than many medieval weapons, but the available evidence is primarily from the very end of the Middle Ages. Only a handful have survived from before the fifteenth century. Historians have thus turned to medieval art to try and fill this void. This approach has provided many valuable insights into the development of the medieval crossbow, but it has its own problems.

Using the study of medieval art as a means the better to understand medieval military technology is hardly a new idea. It has long been common in the study of the medieval longbow where historians have made use of medieval images to support their theories or contradict those of other historians. One school of thought for using medieval art is to examine the work in detail, often using measurements and calculating proportions of figures and other objects, with the goal of extrapolating a medieval reality from its visual representations.[4] This methodology encapsulates a significant problem with using medieval art as an archaeological source. If art is to be used to substitute archaeology, then a judgment must be made about the 'accuracy' of the image. Medieval art generally did not represent the world with high levels of realism. Humans and animals were often distorted and unrealistically proportioned and there was little to no perspective. One could argue that these images were meant to be realistic but were just poorly executed by amateurish artists, but it is not clear that this is the case. It is not only plausible but likely that many works of medieval art are deliberately unrealistic or bizarre, which makes determining which parts of the image may have reflected reality and which did not difficult. For example, in an image of a knight fighting a human-sized snail, maybe the knight's equipment is accurate but given how implausible his foe is we cannot be entirely sure if there are no fantastical elements to his weaponry.

That said, even works that clearly do not represent reality offer the potential of providing insight into the medieval world. These images will not tell us the size of the average medieval crossbow, how deadly it was, or even necessarily what materials were used in making it but that does not mean that they do not offer a valuable insight into its history. They can be used to fill out the chronology of the crossbow, give insight into how it was used by contemporary archers, and even give some insight into its role in broader medieval culture.

Early medieval art

Depictions of the crossbow in early medieval art are few and far between. While the number of depictions might be small in some ways, the early medieval images of the crossbow are among the most important because of how little evidence we have for crossbows during this time. In other periods we have ample textual references to crossbows, and in some cases even surviving weapons, but for the Early Middle Ages we have only these few images and a couple of vague textual references, to be discussed in the next chapter, to cover nearly a millennium of history.

Probably the most famous image of the early medieval crossbow is from a Pictish standing stone found in St Vigeans near Arbroath and dating to between c.871 and c.1000.[5] It is now in the St Vigeans Sculptured Stones Museum in Arbroath. This stone carving includes several hunting scenes on one side, but it is the one at the bottom which shows an archer pulling back the string on what

Detail of hunter on the back of St Vigeans No. 1, St Vigeans Stones and Museum, Arbroath. (Open Government Licence)

appears to be a crossbow that interests us the most. There has been some debate about whether this was intended to depict a true crossbow or if the archer is shooting an arrow guide – a crossbow-like device that did not have a trigger. We can infer that proper crossbows, i.e. ones with triggers, were in use in Scotland at around this period because a bone crossbow nut was excavated near Kilmaurs in Scotland in the nineteenth century. Along with the nut was a coin dating to the sixth or seventh century.[6] While not proof that the weapon on the standing stone was intended to be a crossbow, it certainly provides supporting evidence that makes it seem more likely. If we know that crossbows were used in this region at around this time that should be weighed against the fact that the image is highly abstracted, and it would be hard to argue with certainty that the artist did not intend the weapon to be a crossbow.[7]

Less well known but equally important are two Gallo-Roman stone reliefs now held in the Crozatier Museum in Puy. The first of these images was excavated as part of a Gallo-Roman villa and has been tentatively dated to sometime around the first century AD.[8] The relief shows a party leaving to go on a hunt; two of the hunters near the end of the scene are shown carrying quivers and one of them has a crossbow tucked under his arm. The crossbow's front is pointed forward and its stock is concealed underneath his arm. The bow shows clear signs of recurve, suggesting that it might be of composite design, and is remarkably long in comparison to later medieval crossbow designs – it almost looks like a standard hand bow. This could suggest that this is intended to be a traditional composite bow mounted on a stock for ease of shooting. Certainly in ancient China this was the common practice, so it would not be unreasonable for a Gallo-Roman archer to be using similar technology. It could even be used to support the possibility for the crossbow having been transmitted from ancient China to Roman Europe, although that would be reaching a little too far for just one image to support.[9]

The second image has been tentatively dated to c.300–400 AD, but it was rediscovered in a cellar in 1831 and not in its historic context so that date should

Detail of Gallo-Roman hunting scene in E. Esperandieu, *Recueil general des bas-reliefs de La Gaule romaine* vol. II (Paris, 1908), Item 1683.

be regarded with some scepticism. It is smaller than the previous image and shows the tools a hunter would take with them: a crossbow, quiver, knife or sword, and their trusty hunting dog. This time the image shows the full crossbow from a top-down view. Like the previous image, the bow is very long when compared to later medieval examples, which could further support the idea that crossbows at this time used hand bows instead of specialist bows designed for them. Like with the Pictish stone it is difficult to say whether this was intended to be a full crossbow or an arrow guide since there is no clearly visible trigger mechanism. Certainly, the author does not seem to have intended to depict a rolling nut trigger, but this could be consistent with a simple notch and pin trigger or even one in the style of the bronze triggers of ancient China.[10]

While all three of these images must be taken with the caveat that they could be examples of arrow guides and not true crossbows, they do suggest that European crossbows from this period may have had several significant design features that are absent later in the Middle Ages. The most significant difference is the bows, which seem to be based on handheld composite bows rather than the thicker, shorter and more complex medieval composite crossbow used later. The stocks on these weapons appear to be much shorter than later medieval examples as well, although this could be a stylistic decision by the artist and not reflective of what actual examples of these weapons would have looked like. In total, these three images provide a fascinating if all

Gallo-Roman hunting equipment in E. Esperandieu, *Recueil general des bas-reliefs de La Gaule romaine* vol. II (Paris, 1908), Item 1679.

too limited glimpse into what the crossbow looked like during the late antique and early medieval eras.

High medieval art

By the High Middle Ages, the crossbow was quickly becoming a mainstay of medieval European warfare and the art of the period reflects that growth. Instead of only having a handful of images of crossbows to deconstruct, this period has far more examples than can be covered in this chapter. As a result, this chapter will cover only a selection of the many interesting crossbow-related images dating from between 1000 and 1300.

Plate 10 shows a scene from Beatus of Liébana's book *Commentary on the Apocalypse.* Originally written in the eighth century, this copy of the text was made in 1086 by the scribe Petrus and the painter Martinus and is now held in the Cathedral of Burgo de Osma in Spain. This manuscript includes many fabulous images but the most interesting is a depiction of the Four Horsemen of the Apocalypse. Beatus' version of the four horsemen differs in an important way from the version that is common in modern popular culture. The first horsemen, who in this image is wielding a crossbow in one hand and holding the reigns of his horse in the other, is Jesus Christ. This is made clear by the halo around his head, which usually indicated either a holy figure of at least saintly status, and in Beatus' text there is a clear distinction between the first rider – who is Christ – and three who follow who more closely resemble the modern image of the horsemen.[11]

The weapon in Christ's hand is very clearly a crossbow; the artist has made special effort to show the full trigger system. The bow is a little odd in design, it sort of looks like the two arms are meant to move independently, in a similar manner to how a torsion *ballista* worked, but it seems more likely that it was an artistic choice rather than that hand-portable *ballistae* were used by Spanish cavalry at the time. The stock is short, like those from the early medieval images above, which combined with how Christ is wielding it one-handed raises interesting questions. Archaeological evidence for small handheld crossbows only survives from the very end of the Middle Ages and the early modern era, but this could indicate that the technology was known much earlier. That the weapon is being used by a mounted figure also suggests that the crossbow was possibly used as a cavalry weapon at this time. However, as a depiction of the apocalypse this is not an image that is necessarily entirely rooted in reality. Nevertheless, this is a fascinating depiction of the crossbow that seems to sit between the shorter weapons shown in the Gallo-Roman stones and the larger weapons that will be covered below.

Less than a century later another Spanish artist painted a fresco of a man wielding a crossbow that made up part of the surviving decorations of the Hermitage of San Baudelio de Berlanga in the town of Caltojar in north-west

Spain and is now in the Metropolitan Museum of Art in New York City (Plate 11). This church was once decorated with many frescoes, most of them dating to the first half of the twelfth century. The identity of the original artists is not known and many of the images have suffered significant damage in the intervening centuries. The frescoes include a mixture of biblical stories and secular figures, the crossbowman being an example of the latter – no Jesus wielding a crossbow in this church.

The crossbow in the fresco is huge and appears to have a bow made of wood. It is probably too big given the size of the archer but nothing about the painting suggests that precise realism and scale were a priority for the artist. The weapon clearly has a trigger system so this is definitely a crossbow and not an arrow guide. Given the stylistic nature of the image it would be unwise to draw too many conclusions from it, but it does seem to show a weapon that is more in line with what we would expect of an early wooden crossbow based on our understanding from later archaeological examples and could suggest a continuity of style between high and late medieval wooden crossbows.[12]

The thirteenth-century *Morgan Picture Bible* is one of the Middle Ages' most impressive manuscripts, and one of particular interest to scholars of medieval weaponry. It is a collection of forty-six folios, each of which depicts two scenes from the Old Testament. It was made in the mid-thirteenth century, possibly for King Louis IX of France (r.1226–70).[13] It is alternatively known as the *Maciejowski Bible* which was name of its owner before the Morgan Library acquired most of the folios. In total, forty-three folios are in the Morgan Library, two are in the Bibliothèque nationale de France, and one is in the J. Paul Getty Museum in Los Angeles.[14] Included among the *Morgan Picture Bible*'s folios are several battle scenes from the Old Testament, which include detailed depictions of arms and armour that have fascinated scholars for centuries.[15]

The crossbows included in the *Morgan Picture Bible* are primarily in scenes of sieges, for example folio 10v (Plate 12), Hai Defeats the Israelites,[16] and folio 42r, David Sends a Letter to Joab, displayed on the cover of this book.[17] It is not clear if the crossbows were intended to have wooden or composite bows. The tips of the crossbows seem to be slightly reflexed, that is, they curve forward at the ends, which would usually be indicative of composite bow design. However, reflex is less common in composite crossbows, and it is possible for wooden bows to be reflexed, so either type is still an option. The crossbows all have metal rings on their front which are probably meant to be stirrups, which would be supporting evidence for the widespread use of the stirrup at this time. The circular 'stirrups' look more like hanging rings, used to store the crossbow when not in use rather than to span it, which were common on early modern crossbows. However, given the date of the *Morgan Picture Bible* it seems more likely that these are just unusually circular stirrups, especially when you consider that a crossbowman in David Sends a Letter to Joab clearly has a belt hook, which would have required a stirrup to use.

The crossbows shown in the *Morgan Picture Bible* conform closely to what we would expect of crossbows in this era. We can see solid evidence for the use of stirrups and belt hooks, two technologies that would last the entirety of the Middle Ages, as well as what appears to be Z-triggers and possible even early composite crossbows. The *Morgan Picture Bible* is very detailed and less abstract that some of the works already discussed, but it is still not hyper-realistic and so while it can be indicative of what technologies were in use at the time, we must be careful before we take what it shows as gospel.

Late medieval art

By the Later Middle Ages there are enough surviving crossbows that there is less of a need for historians to use contemporary art to try supplement the archaeological record. However, surviving crossbows are divorced from their original context and through studying the crossbow's depiction in contemporary art, historians can attempt to reconstruct some of that context. In this section we will be examining several noteworthy pieces of late medieval art including an in-depth discussion of one of the most popular subjects of late medieval and early modern art: the martyrdom of St Sebastian.

One fascinating source of crossbow images in the fourteenth century comes in the form of marginalia figures, that is, characters who were drawn in the margins of a book or other document. A truly bizarre set of these can be found in a copy of the *Romance of Alexander* now in the Bodleian Library (Plate 13).[18] This scene, which spans the bottom of a page, depicts human-sized rabbits using crossbows to hunt humans as if they were the prey instead. One figure on the left of the page is taking aim with a crossbow at a human figure while another is spanning its crossbow using a belt hook and stirrup. This is an amazing image because it both captures the sometimes utterly surreal nature of medieval art while also providing great insights into how crossbows were used. We can see a clear illustration of the principles involved in using a belt hook as well as a figure, albeit one who technically does not have hands, taking careful aim with a crossbow. No single image better encapsulates how medieval art can provide insights into contemporary behaviour while also being completely unrealistic.

A similarly intriguing and bizarre scene can be found in the *Coëtivy Hours*, an illuminated book of hours made between 1443 and 1445, and now in the Chester Beatty Library in Dublin. Plate 14 shows a full page from this book featuring a dancing skeleton, a literal embodiment of the late medieval concept of the *Danse Macabre*, holding a giant bolt – possibly an oversized crossbow bolt or else one for use in a springald or great crossbow.[19] In the margins at the bottom of the page an archer with a crossbow whose torso emerges from a large flower is taking aim at a bird. Again, the crossbow is detailed with a stirrup that the archer would struggle to use on account of his lack of legs. This image also shows how crossbows and their paraphernalia could feature in aspects of medieval life beyond just the purely military and could have an additional symbolic meaning.

One of the most iconic images of medieval warfare, and one which adorns the covers of more than a few histories of the Hundred Years War, is an illumination of the Battle of Crécy from a fifteenth-century copy of Froissart's *Chronicles*.[20] This image (Plate 15) is also one of the best medieval examples of the operation of a windlass. In the foreground a Genoese crossbowman has been drawn in the action of spanning his weapon with his windlass. His foot is in the stirrup to steady the crossbow as he cranks the handles. The image clearly illustrates why crossbows spanned with a windlass needed long stocks since it allows the archer to perform this operation while standing upright. The relative size of the windlass is remarkable, and it captures how even though they were not very heavy their bulk would still have made them difficult to handle. This is reinforced by the fact that the artist has not bothered to draw a windlass for any of the other crossbowmen, who are either shooting or fleeing the battle. This image presents a beautiful example of how this iconic spanning device was used as well as showing how it endured throughout the Middle Ages. The windlass was first introduced over a century before the Battle of Crécy and at least half a century later it was still being depicted in medieval art.

The Battle of La Higueruela, fought in 1431 by the Kingdom of Castile and the Sultan of Granada, was not one of the most significant or famous battles of the Middle Ages. However, thanks to a late sixteenth-century fresco in the Hall of Battles in the monastery of El Escorial in Spain it is one of the most magnificently memorialized (Plate 16).[21] This is a fascinating source because it is an early modern artist depicting a then largely obsolete weapon being used in a battle approximately 150 years earlier. The version of the battle is a romantic rather than realistic one but nevertheless, as should be clear by now, that does not mean it cannot be informative. For the purposes of this discussion, we will be focusing on a small section of the overall fresco which shows several crossbowmen, weapons resting against their shoulders, as well as a workman using a spanning bench to restring a crossbow. The crossbows all look to be wooden, which is interesting because while wooden crossbows would still have been used in 1431, they had pretty much disappeared by the mid-sixteenth century. This could suggest some institutional memory of the wooden crossbow, but it could also be a misinterpretation on the part of the artist or by the modern viewer examining their work. The overall designs of the weapon are simple but evocative, with their short stocks, clear Z-triggers and stirrups attached to the stocks with a mixture of lashings and metal plates.

The inclusion of the men operating a spanning bench shows an aspect of medieval warfare rarely mentioned in other sources: the artisans who accompanied medieval armies. The most famous of these were the engineers responsible for building and operating siege weapons such as trebuchets or siege towers, but equally important were blacksmiths or, in this case, crossbow makers who could be relied upon to perform maintenance on weapons as needed. While it is a little odd that this man is restringing a crossbow so close the main action of the battle, it does represent a job that someone would have had to do while an army was on campaign. Unlike a longbow, restringing a heavy crossbow was an involved process that required specialist equipment. If an army neglected this, they ran the risk of expensive weapons being rendered useless by an unfortunate accident. This image captures an often-overlooked element while also providing a fascinating image of what restringing a crossbow looked like with pre-modern equipment.

St Sebastian

St Sebastian was a Christian martyr from the third century who was tied to a post and shot full of arrows by Roman soldiers at the orders of the Emperor Diocletian (r.284–305) as part of that emperor's persecution of Christians. Starting in the 1400s St Sebastian became a very popular figure for artistic depictions and hundreds of paintings and sculptures of him survive from the fifteenth and sixteenth centuries. He is, among other things, one of the patron saints of archers, but his role as a patron saint of the plague-stricken probably better explains his popularity during the late medieval and early modern periods. While most depictions of the saint show him tied to the post and with arrows protruding from his mostly naked body, this moment was not his actual martyrdom. After he passed out St Sebastian was left to die by the Roman soldiers, but he was rescued and nurtured back to health by the widow St Irene of Rome. After his recovery Sebastian sought out and harangued Emperor Diocletian for his persecution of Christians and the emperor, astonished to see him still alive, ordered that he be beaten to death right then and there.

The martyrdom of St Sebastian is an invaluable source for images of crossbows in the later Middle Ages. The detail on these works is often very impressive. Take for example the woodcut of the Martyrdom made by Master ES c.1450 and now in the National Gallery of Art in Washington, DC.[22] This image includes three archers all with a similar style of crossbow, two who are taking aim at St Sebastian while another is busy reloading his weapon. The archer in the foreground is holding the bolt on his crossbow in place with the thumb of his left hand, a common practice before the invention of

Master ES, Martyrdom of St Sebastian, c.1450/60, National Gallery, Washington DC.

the bolt clip. This serves the obvious practical purpose of keeping the bolt steady while the archer lines up their shot. This technique is more often associated with shooting down from an elevated position, such as from a castle or town wall, but this image shows that it could be used when shooting on a flat surface as well. The archer in the middle ground is spanning his weapon in an interesting way. It looks like he is supposed to be using a *krihake* but his crossbow completely obscures the mechanism. He is shown sitting down, so either he is meant to be spanning his crossbow while sitting or, given that the string already seems to be pulled back to the lock, he may be finished and now unhooking the spanning device.

The third archer in the background seems to be aiming out at the viewer rather than at St Sebastian. The image of a crossbowmen aiming at the viewer was something of a motif in early modern woodcuts, although usually it is independent of the context of St Sebastian. There is a similar image attributed to Cherubina Alberti, dated 1579, in the same gallery as well as a famous image of an archer and a milkmaid from c.1610 now in the Metropolitan Museum of Art.[23] Perhaps most sinister of all is the late seventeenth-century engraving of a skeleton in pseudo-classical attire shooting out at the viewer, now in the Wellcome Collection in London.[24]

A fifteenth-century Martyrdom of St Sebastian by Master der Heiligen Sippe includes an archer spanning his crossbow with a *cranequin* (Plate 17).[25] His crossbow is placed on the ground, we cannot see if he has a stirrup, and the *cranequin* is facing towards him as he operates it – an interested detail for how *cranequins* were used that cannot be gleaned by the device alone. Based on its colour and dimensions it appears to be a steel crossbow, and it is significantly larger that the crossbows shown in the engraving by Master ES, indicating that crossbows in late medieval Germany probably came in a range of sizes. The length of the stock is so long that it looks like it would be better suited to being spanned by a windlass. That it is being spanned by a *cranequin* shows that the type of spanning device cannot be assumed by the shape of the weapon alone.

Piero de Pollaiuolo's painting of the Martyrdom of St Sebastian from c.1475 is one of the most impressive images of the saint ever made (see Plate 18). This massive painting now takes up most of a wall in London's National Gallery.[26] Two of the archers in the painting, both standing in the foreground, are spanning their steel crossbows with what appear to be *krihakes*. Of particular interest is the placement of the crossbow with the bow passed between the archer's legs and probably pressed up against their calf muscles. Most modern interpretations of how to use a *krihake* assume the crossbow was placed straight ahead of the archer, like the crossbow being spanned by a *cranequin* in Master der Heiligen Sippe's painting. Instead, this painting suggests a very different stance, one closer to a squat. It would be reckless

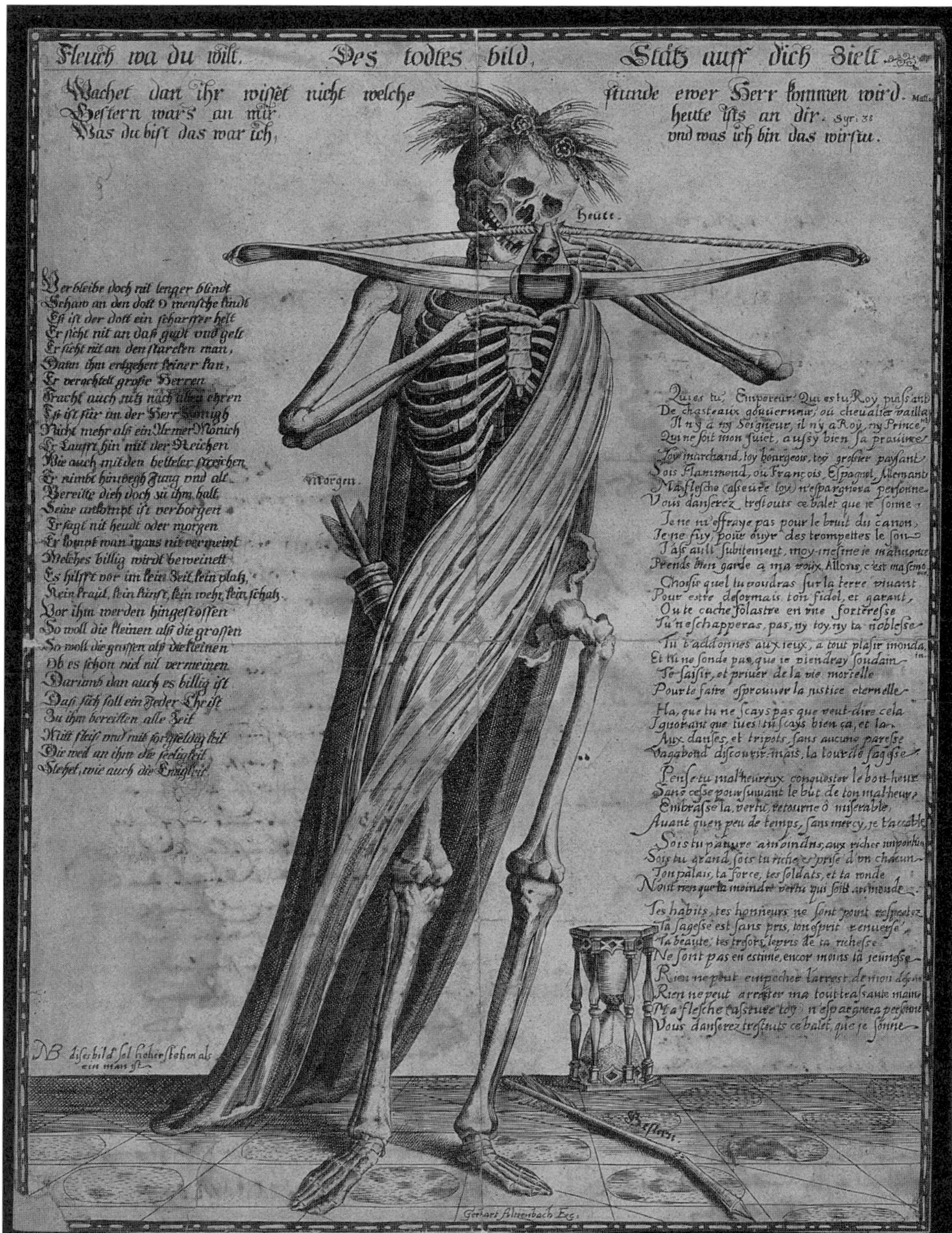

A skeleton with bow and arrow. Etching attributed to Gerhart Altzenbach, seventeenth century. Wellcome Library, London (Public Domain Mark). (Designated CC0 Public Domain)

to assume that the method portrayed by Piero de Pollaiuolo was the only way this device was used, but it is an excellent insight into the methods for spanning the crossbow.[27]

The Martyrdom of St Sebastian remained a popular subject into the early sixteenth century with a remarkable continuity in style. A famous triptych by Hans Holbein the Elder from c.1516 depicts the martyrdom in the central panel (see Plate 19).[28] This image includes three crossbowmen in total, two wielding steel crossbows and one with what appears to be a composite crossbow. All three crossbows are very small, close in size to those in the Master ES woodcut. The steel crossbows have bolt clips, a useful feature for holding bolts in place and the origin of which is still much debated. The image also shows a variety of bolt types: the two archers on the left seem to have military-style bolts while the one on the right has what appears to be a crescent-shaped hunting bolt. There is a lot of little detail in this image that presents intriguing glimpses into what contemporary crossbow design and use would have been like. The crossbowman in the foreground is kneeling down to span his crossbow with a *cranequin* with a bolt clenched in his teeth, a fascinating little detail about how archers could keep their ammunition to hand.

Pietro Perugino completed his fresco of the Martyrdom of St Sebastian in 1505 (see Plate 20).[29] His painting includes an interesting example of an archer spanning his crossbow. The stirrup is resting against his thigh while he has both hands on the far end of the stock. He appears to be about to span the crossbow with a goats-foot lever, but the details are hard to distinguish. The method of bracing the crossbow stirrup first against his leg is not how the use of the goats-foot lever is usually depicted, although this is not the only example of this technique. It could be that this was a style distinct to the region of the artist, or it could be that this was just a less popular method, so we only have a few examples showing it. It is yet another example of how artistic images present a more varied and intriguing past than could be inferred from the archaeological record alone.[30]

Crossbow Making in Art

Medieval art also has the potential to provide insight into the context in which medieval crossbows were made. The making of medieval crossbows has unfortunately been a somewhat neglected area of research. There are plenty of references to crossbow makers – nearly every town in the Holy Roman Empire large enough to have walls had at least one and some large towns had many – but it is an area that is only now beginning to be explored in the detail it deserves. There are also several images of crossbow makers working in their shops that can give us a glimpse into their world. Unfortunately, these all come from the sixteenth century – when the decline of the crossbow as a military weapon rapidly shrank the available market for crossbow makers, putting the profession into a decline as well. Some crossbow makers persisted and were able to live off of supplying weapons to shooting guilds and hunters, but the number was much diminished from previous centuries. Still, it is worth studying these images as they have the potential to provide insight into both the means of crafting a crossbow as well as the lives of those who did the crafting.

Olaus Magnus' (1490–1557) *Historia de Gentibus Septentrionalibus*, or the *History of the Nordic Peoples*, includes an excellent woodcut of a crossbow maker.[31] This whole book is fascinating and provides plenty of insight and information about contemporary Nordic culture – albeit from something of an outsider and at times flawed perspective. It is also filled with dozens of woodcut vignettes, several more of which will appear elsewhere in this book. The image being discussed here shows a master craftsman finishing off a steel crossbow held out in front of him while two of his previously-completed works hang above his head. On the table beneath the crossbow, we can see the tools of his trade: some are in active use as he finishes the weapon while others are awaiting his next project. Beyond the crossbow we can see a belt hook and what appears to be a windlass or possibly a *krihake* – both clearly still popular methods of spanning crossbows in the mid-sixteenth century. On a rack in the back of his workshop is arrayed a variety of bolts with different types of heads. The vast majority, very possibly all of them in fact, are intended for hunting – each suited to a different kind of prey. While we are left to infer certain aspects of what he is doing, especially since the weapon seems nearly finished, this image is a fascinating glimpse into the workings of an early modern crossbow maker's shop. One can only wonder how similar it would

Crossbow Maker in Olaus Magnus, *Historia de Gentibus Septentrionalibus* (Rome, 1555).

have been to what his grandfather or great-grandfather might have been doing a century earlier.

Hans Sachs and Jost Amman's *Book of Trades*, published in Frankfurt am Main in 1568, was a series of 133 woodcut illustrations accompanied by witty poetry, showing a variety of trades ranging from the Holy Roman Emperor himself to the humble crossbow maker.[32] The latter shows the crossbow maker working on a nearly-finished crossbow stock but without any of its extra pieces – no bow or trigger yet. Arrayed on the counter in front of him are his tools. He is working in front of an open window where someone is leaning in, presumably engaging him in conversation. Hanging above the counter are two finished crossbows as well as a bundle of bolts. In the background is the edge of what looks like a stringing bench. As with the previous image, this provides a very intimate picture of what a crossbow maker's workshop looked like at the time. Where the previous one showed a lone craftsman working in isolation,

Crossbow maker in Hans Sachs and Jost Amman, *Das Ständebuch* (Frankfurt am Main, 1568). Quire L6. (Public Domain Mark, Wikimedia Commons)

this woodcut shows someone engaged with his community, which captures how crossbow makers were also often major figures in their civic communities.[33]

The Crossbow as Art

Depictions of medieval crossbows in art are all well and good, but what about the crossbows themselves? Many surviving medieval crossbows are now kept in art museums, such as the Wallace Collection in London or the Metropolitan Museum of Art in New York City (the Met), which reflects their status as art objects in addition to being weapons. This section will explore how medieval crossbows were works of art in their own right in addition to being weapons.

The most common decoration for medieval crossbows was the addition of inlays in the stock. These were usually made from staghorn but could be made from animal bone, other types of horn, or in rare cases ivory. A standard example of this type of decoration can be found in a crossbow now held the Met.[34] This crossbow (Plate 21) has a composite bow and dates from the middle of the fifteenth century. It has a simple hanging ring mounted on the front, and halfway along the stock a hook possibly for use with a *krihake*. It has several separate decorative horn inlays – one along the top, two on each side and one along the bottom. The two on the sides include one inlay spanning most of the length of the stock as well as a smaller piece around the hole where the rope bindings pass through it. The inlays are relatively simple geometric shapes except for two places on each of the sides where they form an impressive wavy design. It is an elegant and somewhat minimalist design – it is beautiful without being ostentatious.

The crossbow of Count Ulrich V of Württemberg (1413–80) is roughly contemporary with the crossbow discussed above and also in the Met's collection (Plate 22). The maker's mark on it indicates that it was probably made in Switzerland between 1453 and 1460. This is also a composite crossbow but unlike the previous example it is lacking any clear spanning aids. The decoration has a similar style the previous example, but it includes much more detail. In addition to the horn inlays there is significant decoration on the wood of the stock. These are elaborate swirling patterns that must have been painstakingly carved by a master artisan. The horn inlays are fewer than in the previous crossbow but where those were plain several of the ones on Count Ulrich's crossbow have intricate carved patterns. Above the trigger is carved a German benediction rendered in Hebrew letters and heraldry is carved on either side of the stock – on the right is the heraldry of Württemberg while on the left is that of Savoy. The direct association with a historical figure as well as the elaborate decorations make this one of the most impressive surviving medieval crossbows – you can see that it is the work of highly skilled artisans who created a majestic piece of art in addition to a weapon for killing.[35]

Moving from counts to kings, the next crossbow once belonged to Matthias Corvinus, King of Hungary from 1458 to 1490. It was probably made in the year 1489 (Plate 23). Unlike the previous two examples this is a steel crossbow, but it is similarly decorated with staghorn inlays. Like the crossbow of Count Ulrich this crossbow has highly ornate carvings decorating the wood of the stock, but it is in the horn inlays that it truly shines. The elaborately carved horn sections make up a much larger proportion of the decorations than they did on Count Ulrich's weapon. Matthais' crossbow is decorated with detailed carvings of dragons on one side and an equally detailed depiction of St George on the other. It also features both Matthias' personal arms as well as those of the Kingdom of Hungary. The scale of the detail on this weapon is staggering, and it must have taken very skilled artisans a long time to create each and every image and component of this weapon.[36]

One thing we must caution against is an assumption that because these three weapons examined above are in an approximate chronological order from oldest to youngest that these decorations should be seen as changes in artistic style over time. Instead, they should probably be seen as three different tiers of decoration. The first case is a decorated crossbow of unknown provenance which may indicate that it was a weapon for a lesser noble or even potentially a wealthy merchant or town burgher. In contrast, the crossbows of Ulrich and Matthias are weapons made for a count and a king respectively – and the decorations reflect those levels of status. This is not to say that only a king could own a weapon as highly decorated as Matthias' crossbow, but rather that in crossbow decorations – as with many things – you got what you could afford to pay for. These three weapons should be seen as the range of what crossbow makers in central Europe were capable of in the fifteenth century.

Another slightly rarer style of crossbow design consisted of covering the entire stock in horn. These could take the form of a plain undecorated layer of stag horn or bone, see for example crossbow XI.104 now in the Royal Armouries in Leeds (Plate 24), or examples where the horn was decorated with engravings, such as Wallace Collection A1032. Royal Armouries XI.104 is a composite crossbow, probably made in Germany c.1500, and its stock is entirely covered in undecorated white bone. The plain colour and simple decoration are striking. In contrast, the stock of A1032 features a plethora of figures carved in relief across most of the stock. The only place free of carvings is the section that the archer would have placed against themselves when taking a shot – presumably because they would have made the weapon uncomfortable to use, showing that despite its decoration it was not just intended for display. The stock displays numerous figures engaging in a variety of activities as well as a range of coats of arms – the ones on the side seem to have been carved and then painted while the one on top of the stock near where the bow attaches to the stock just seems to be painted on.

A word of caution about studying surviving medieval crossbows and the pitfalls that can arise from it: it must never be forgotten that these weapons were not magically transported from their original owners and into the collections of modern museums: they have had long and sometimes quite complex modern histories as they passed between various owners and collectors. An excellent example of the pitfalls that can arise from this is found in another crossbow now in the Met's collection. This weapon has generally been dated to either the late fifteenth century or possibly the early sixteenth century, but the museum's curators admit that it could be from even later. It is a beautiful steel crossbow with a highly decorated stock covered in carved stag horn. The carvings depict scenes of medieval or early modern hunters amidst scenic backdrops – a truly wondrous piece of art. Carved on the stock in clear numerals is the date '1584'. However, this date was added to the crossbow in the nineteenth century and its addition must raise some questions about the rest of the decorations. Are these truly authentic medieval or early modern carvings or were they added to the weapon in the modern era? Or are only parts of the decoration later additions, and if so which parts? Modifications to medieval crossbows were all too common and historians must be cautious when examining these weapons.[37]

Not every crossbow needs to be ostentatious or elaborate in its decoration: there is a beauty to be found in simplicity as well. Nowhere is this more apparent than in the Spanish-style crossbows that emerged at the end of the fifteenth century and were very popular during the sixteenth century. These weapons were always made with steel bows and had thin stocks. While they originated in Spain, they could be made in other parts of Europe. They varied significantly in size, with some being over a metre long while others were half that length. The collection in Grandson Castle has great examples of both the large and small types.[38] A fabulous mid-sixteenth century example is now in the Art Institute of Chicago (Plate 25).[39] It has several simple brass plates to reinforce the lock and the bow, but otherwise is entirely undecorated. The attraction of this weapon is all in its sleek shape. Just like how you cannot help but admire a well-fashioned rapier, this weapon captures a simple but deadly energy in its sleekness.

This section has hopefully made clear that the creation of a medieval crossbow was not necessarily just the fashioning of a weapon but also a chance for artistic expression – both from the maker and from whoever commissioned it. It can be tempting to assume that all decorated crossbows were owned by nobles and societal elites and were exclusively used for recreation, but it may not be so simple. Many surviving crossbows have simple horn decorations and the possibility that some of these weapons would have been used for war cannot be ruled out. Just as swords, shields and plate armour were decorated despite their intended martial purpose, so war crossbows probably had some decoration. Artistic expression and warfare are not mutually exclusive, and the crossbow must be viewed holistically as both a work of art and a tool of death.

Part II:
The Crossbow – A History

Chapter 4
The Origins of the Crossbow

King Edmund II 'Ironside' (r.1016) is not a particularly famous English monarch. He was the son of the infamous Æthelred II 'the Unready' (r.978–1013) and ruled for less than a year, from April to November 1016. In that time, he fought at least four battles in an attempt to drive off an invasion of Danes led by Cnut, the future king of England, Norway and Denmark (r.1016–35, Norway from 1018, Denmark from 1028). In terms of ratio of battles fought to time on the throne Edmund might hold the record for an English king, even if he did not have as good a record for winning those battles. It is not his brief reign that is of interest here, however, instead it is the circumstances of his death on 30 November 1016.

The chronicler Geoffrey Gaimar (fl. 1130s) wrote an interesting account of Edmund's death in his *History of the English People.* This is one of the oldest historical chronicles written in French and was composed in lines of eight-syllable rhyming couplets sometime between 1136 and 1140. In constructing his narrative Gaimar likely drew on a range of sources, being also known as a translator of Old English and Latin. His account of Edmund's death is worth quoting at length here, as it is truly fascinating:

> Eadric caused an engine to be made,
>
> The bow which would not miss, he knew how to draw.
>
> If anything touched the string,
>
> Soon would one hear bad news.
>
> ...
>
> Where this bow was prepared
>
> He had placed a new house.
>
> Privy house they called it,
>
> Men went there for that purpose.
>
> The king was taken there at night,
>
> As Eadric had commanded.

Directly he sat on the seat
The arrow struck him in the fundament.
It went up as far as his lungs.
The feather never showed
Of the arrow which was in his body,
And no blood came forth.
The king cried a death cry,
The soul fled from him, he was no more.[1]

The word crossbow is never used, but the description sounds a lot like a trap bow, a special type of hunting crossbow, although one usually associated with a later time period. Gaimar added that the treacherous Eadric went to Cnut in hopes of receiving a reward for killing Edmund, but Cnut rewarded Eadric by beheading him with an axe. If this story is true, then Edmund could potentially be the earliest European killed with a crossbow that we can put a name to. Unfortunately, history is never that simple and Gaimar's version of events was contradicted by other authors.[2]

Henry of Huntingdon wrote his *Historia Anglorum* at around the same time as Gaimar and included his own version of the death of Edmund Ironside. In Henry's version Edmund still died ignobly on the toilet and his killer was still beheaded by Cnut, but instead of an elaborate trap Eadric just stabbed the king to death – a far more popular method of medieval assassination.[3]

However, both accounts were written over a century after Edmund's death and earlier authors did not describe Edmund as dying on his toilet. The *Encomium Emmae Reginae*, an eleventh-century Latin chronicle written in honour of Emma of Normandy, Edmund's stepmother and later the wife of King Cnut, does not describe how Edmund died, but it does still include the story of Eadric's beheading by Cnut, although in this case it is because Eadric was a coward who had refused to fight with his lord Edmund against the Danes and later tried to spin this as a positive to Cnut after Edmund's death.[4] Historians have generally accepted the testimony of the *Encomium* and other contemporary authors who do not provide a salacious story of Edmund's death – the assumption being that he either died of natural causes or from wounds sustained in one of his many battles.

Edmund was not the first European killed by a crossbow, of that we can be certain, but what do we know about the early history of the crossbow in Europe? It is impossible to determine exactly when the crossbow was first used in Europe. The archaeological record is extremely sparse before the Later Middle Ages and textual references are few and far between. Before examining the history of the European crossbow, it is important to consider where the crossbow was first used.

Ancient China

The list of technologies with their origins in China is extensive and so it should not be a surprise to find the crossbow among them. In a rarity in the history of crossbows, not only is there evidence of crossbows in ancient China, but there is also an abundance of evidence. The oldest complete crossbow was discovered as part of recent excavations of the famous Terracotta Army of Emperor Qin Shi Huang (259 BC–210 BC), meaning it must date from either 210 or 209 BC. In addition to the complete crossbow, numerous bronze triggers have been found among the Terracotta soldiers. A study of these crossbows shows a remarkable level of standardization in their manufacture, suggesting that they may have been mass produced. This trait is also apparent in the crossbow triggers that survive from the Han Dynasty (202 BC–220 AD). The bronze trigger system also proved very popular outside of China and spread to Korea and south-east Asia despite attempts by the Han to ban its export.[5]

The bronze crossbow triggers from this period worked very differently to the rolling nut trigger found in medieval Europe (see Plate 26 for an example of a Han dynasty trigger). The Chinese triggers worked a bit like a door latch, there was a lever at the bottom which when pushed forward lifted two claws out from the trigger block which would catch the string. When the lever was pulled the piece that supported the claws would drop away allowing them to fall back into the trigger block, releasing the string. Ancient Chinese crossbows had several major differences from their medieval European counterparts beyond the trigger mechanism. The bows of these crossbows were standard composite bows of the era, the same kind that a traditional archer would use. These were mounted on to wooden stocks with the trigger block placed at the very end of the stock. Because the bows were hand bows, to get their full power they had to be drawn at least 26in, meaning that the draw of the crossbow was the full length of the stock. In contrast, medieval European crossbows usually had a draw distance between 7in and 10in – usually less than halfway down the stock.[6]

There is evidence of crossbow use in China even earlier than the surviving triggers, which likely represent the technology already in a mature form. The oldest reference to a crossbow comes from Sun Tzu's *The Art of War*, which as a text is difficult to date but could be from as early as 498 BC. There are several references to crossbows in Sun Tzu's book, for example one passage goes 'Their strategic configuration of power [*shih*] is like a fully drawn crossbow, their constraints like the release of the trigger'.[7] The fact that Sun Tzu uses the crossbow as a metaphor suggests that his intended audience was already intimately familiar with the weapon, meaning that it probably was not a new technology.

Joseph Needham has argued that based on the available evidence, including early linguistic evidence, the crossbow might very well have its origins in societies that existed before the rise of what historians would recognize as a coherent

Chinese culture. However, he also noted that in the earliest Chinese sources they are very clear that the crossbow originated within China. Some Han texts created a mythic story of a man named Chhen Yin who was from the nearby state of Chhu. Chhen Yin supposedly introduced the crossbow to the Han court but he himself was not the inventor of the weapon, instead he credited a series of masters in Chhu's recent past, starting with a Mr Chhin. The technology was passed down before it was taught to Chhen Yin and then in turn passed to the Han. It is an interesting story and possibly could represent a grain of truth that the Han acquired the crossbow from a neighbouring people, but by the time the Han came to rule China the crossbow had already been an established part of Chinese warfare for centuries.[8]

While its origins are shrouded in mystery, historians can be relatively certain that the crossbow was relatively well established during the Warring States period (476 BC–221 BC) and was widespread by the fourth century BC. Han Dynasty armies used the crossbow almost exclusively, preferring it to the traditional bow. It helped the crossbow's popularity that the traditional bow was associated with the nomadic enemies that the early Chinese states were often at war with. The crossbow remained a popular weapon in Chinese warfare for centuries after the fall of the Han. Subsequent dynasties tried to restrict access to the crossbow, with mixed success. The Song Dynasty (960–1279 AD) relented and made more allowances for its sporting use but it remained primarily a weapon of state-sanctioned violence. In the mid-thirteenth century the Franciscan diplomat John of Plano Carpini (c.1185–1252) wrote while visiting the Mongol court in China that if anyone wished to fight the 'Tartars' they ought to use the crossbow, declaring that they were very much afraid of it.[9]

Is the European crossbow derived from the Chinese one or is it an independent invention? While the crossbow dates at least as far back as the fifth or sixth century BC, the first solid evidence for the crossbow in Europe is at least 700 years later, and archaeological evidence first survives from several centuries after that. There are no surviving bronze triggers like those in China in pre-medieval or medieval Europe. When crossbows did become common in Europe during the Middle Ages the design of the weapons was radically different from the Chinese originals.

Joseph Needham proposed the possibility of multiple waves of Chinese technological influence, one in ancient antiquity through direct Roman and Han Dynasty contact and then a second one in the Middle Ages via the Islamic rulers of the Middle East. This explanation is based on the now somewhat outdated belief that the crossbow was used by the Romans and then disappeared until the ninth or tenth centuries AD. However, Needham's observation that the initial references to crossbows in Roman sources aligns with the first sustained contact between Rome and China is hard to dismiss. That still does not explain the lack of any surviving Chinese-style triggers from Rome which when combined with the overall dearth of evidence on what Roman crossbows were like makes it hard to prove this link definitively. While the most likely explanation is still that the Chinese crossbow underwent substantial changes to eventually become the European crossbow, there is a possible alternative heritage for the European crossbow. This one comes from another source of many early human inventions: ancient Greece.[10]

Ancient Greece

The other possible point of origin for the European crossbow comes from an obscure Greek weapon known as the *gastraphetes*. It consisted of a large bow mounted on a stock, very similar to a crossbow, but what made the weapon unusual was its method of spanning. The *gastraphetes* was spanned using a wooden bar that extended from the front of the weapon. The archer would place the bar against the ground and then with their stomach pressed against the end of the stock they would use their body weight to drive the bar back into the stock which would push the string into the trigger system – which was a claw-like mechanism distinct from both the ancient Chinese trigger and those found in later European weapons. The use of the archer's stomach during the spanning process gave the bow its name, *gastraphetes* meaning 'belly releaser'.

No parts of a *gastraphetes* have survived, let alone a complete weapon, so there is no archaeological evidence to examine. Instead, historians are reliant upon the writings of the mathematician and engineer Heron of Alexandria (c.10 AD–c.70 AD). A tenth-century copy of Heron's work *Belopoeica* included a detailed diagram of how a *gastraphetes* was supposedly constructed (Plate 27). Some doubts must remain about whether the tenth-century illustrator was familiar enough with the weapon to draw an accurate diagram, or whether they were working from an older version that has not survived. Heron himself was working from the works of the earlier Greek engineer Ctesibius (c.285 BC–222 BC), which could suggest an earlier origin for the weapon. Heron believed the *gastraphetes* to be a precursor to the catapult, which could put its origins sometime in the fifth century BC, but that is a claim from an author writing half a millennium later so it should be viewed with healthy scepticism.[11]

It is impossible to say with any certainty the extent to which the *gastraphetes* had an impact on the design or development of the medieval crossbow. It was a niche weapon with unique spanning and trigger mechanisms, but at the same time medieval crossbows do not much resemble Chinese crossbows either. The inexact dating of the *gastraphetes* raises further problems. If Heron was correct and it pre-dated the catapult than it seems likely that it was itself an independent invention, but if it was from much closer to Heron's own lifetime then historians must consider the possibility that the *gastraphetes* could have derived from Chinese crossbows. The *gastraphetes* could potentially be the origin of the

European crossbow, but it is also possible that it was simply an interesting dead end in the history of technology.[12]

Ancient Rome

Whether ancient Romans used the crossbow or not has been a subject of extensive scholarly debate over the past century. This revolves around the exact interpretation of several somewhat obscure Latin words as well as the reliability of the sources using them. The scholarship on this subject is extensive and relies in many parts upon a significant knowledge of both classical Latin and how it was used by a handful of Roman authors. The version of this debate presented below will necessarily be abridged and simplified as it is beyond the scope of this book to relitigate Roman history.[13]

The discussion of the Roman crossbow hinges on the interpretation of the term *ballista* and several terms derived from it. As anyone familiar with ancient Roman siege weapons will know, the *ballista* was a torsion weapon with two horizontal arms joined by a single string. Torsion weapons are powered by twisted rope or sinew that generates energy when the arms of the weapon are pulled back, the catapult being the most famous example. It is worth noting that some scholars do not distinguish between catapults and *ballista*, instead classifying them as different types of *ballista*. For the purposes of this book the term *ballista* will be used to refer to the two-armed variety of this weapon. While the *ballista* resembled a large crossbow and both weapons could shoot bolts or stones, the mechanics that powered them were very different and the *ballista* should not be seen as an early type of crossbow.

In his influential manual on the practice of warfare the late Roman writer Vegetius mentioned two specific types of *ballista*: the *manuballista* and the *arcuballista*. The first approximately translates to hand-*ballista* while the second has generally been translated to bow-*ballista*. Both of these terms sound like Vegetius might have been referring to handheld weapons that resembled a *ballista* or, in other words, a crossbow. Vegetius offered little detail or clarity since the two terms were used only sparingly in his writing, leaving historians to try and interpret what he meant. Was he describing a new weapon or was this something that had existed for centuries and long been a staple of Roman warfare? Does the only-passing reference to the weapon indicate a general lack of interest in them on the part of all Romans or does it reflect Vegetius' own bias? It is worth noting as well that while many historians have accepted that these terms referred to a type of crossbow, others have argued that what Vegetius was referring to was instead a type of torsion-powered *ballista*. The vagueness of Vegetius' writing certainly leaves open the possibility that the latter could be true. The term *arcuballista* seems to be unique to Vegetius and authors who based their own writings on him while *manuballista* also appeared as a term in the anonymous *De Rebus Bellicus*. The anonymous author primarily described torsion-powered weapons in their text which could suggest that they meant *manuballista* to be a type of torsion weapon, but there is also no reason to assume that if he meant a torsion weapon that must mean that Vegetius did as well.[14]

The sparse written record has been supplemented by an equally sparse archaeological one. A handful of crossbow nuts have been found in excavations that have been generally dated to the Roman period, although it has not been possible to date them with much exactness, not even necessarily to a specific century. These finds, along with the Gallo-Roman carvings discussed in Chapter 3, certainly provide supporting evidence for the use of the crossbow in Europe during the first millennium AD but they do not provide much in the way of specifics for historians to work with.[15]

The crossbow was probably in use in Europe at least from the late Roman period and possibly earlier but it is impossible to say with any certainty how widespread it was. The European crossbow may owe its invention to the *gastraphetes*, but the evidence for the ancient Greek weapon is very thin. A more likely explanation is that the European crossbow's origins lie in the late Roman period, with a very plausible explanation being that it was the result of greater contact between Rome and ancient China.[16]

The Crossbow in the Early Middle Ages (c.500–c.1000)

The earliest solid evidence for the use of crossbows in medieval Europe, two Gallo-Roman stone reliefs and a Pictish one, have already been discussed in Chapter 3. These images present strong evidence and provide some tantalizing hints but they leave historians wanting to know more, like what kind of triggers or bows these crossbows were made with. They also show that the crossbow was used across a wide geographic area, from southern France all the way to Scotland. All three images happen to be of hunting scenes, showing that the crossbow's popularity with hunters started very early as well. Unfortunately, three stone relief images are not a lot of evidence to go on and for several centuries that is pretty much all the evidence that historians have. It seems likely that crossbows were also used in war at this time, but there is little definitive evidence to support it. Bernard Bachrach has made a strong argument that crossbows were probably used by the armies of the Carolingian Empire (613 AD–877 AD), but the first definitive evidence for the crossbow's use in medieval warfare comes in the ninth century.[17]

On 24 November 885 a Viking army from Denmark arrived outside the city of Paris. They had sailed up the Seine as far as Rouen and then marched overland the rest of the way. Paris still had its original Roman walls protecting the Île de la Cité but had long since expanded beyond just that island. To protect this greater urban area two fortified bridges had been built on the river. The city was defended by Count Odo of Paris (c.857–98) who would later go on to become the King of West Francia in 888. After an initial unsuccessful assault immediately upon arriving, the Vikings settled in for a more traditional siege. They built a fortified camp near the city to protect themselves against counter-attacks and began constructing various siege weapons. What followed was several months of Viking attacks utilizing their new equipment, including battering rams and an attempt to send flaming ships to ignite the bridges. The French successfully drove off each attack in turn. In March 886, after three months of near-constant siege warfare, a relief army came to the aid of the city and the Vikings withdrew after negotiating for the payment of sixty pounds of silver.[18]

A Benedictine monk named Abbo Cernuus (fl. ninth century) happened to be present in Paris for the Danish siege and over a decade later he wrote a detailed account of it. Unfortunately for historians hoping for a clear narrative of events, Abbo wrote his account as a verse poem filled with classical allusions and exaggerated details. While the poem makes for a truly engaging read, constructing a coherent narrative of the siege from it has long been a challenge for historians. Abbo described as Viking army of 40,000 warriors on 700 ships, an obvious exaggeration, as no such force would have been able to survive on the available food during winter around Paris. Abbo also described a variety of weapons used in the siege, which could include one of the earliest references to the use of the crossbow in medieval warfare. Abbo made several references to *ballista*, and from context he probably meant the torsion-powered siege weapon and not the crossbow, but in one passage Abbo wrote of the Danes: 'They sharpen, mend, and hammer out swift-flying bolts.'[19] This could be a reference to ammunition for the *ballista*, but it could also mean bolts for crossbows. In several passages he also described darts being shot with deadly results, which could also possibly mean crossbow bolts. The mixture of references to contemporary European and classical Greek and Roman weapons makes interpreting Abbo a challenge and raises many questions about his reliability, but there is certainly a valid interpretation of his work that suggests he was describing crossbows at the siege – but there is also an equally valid interpretation that he was not.[20]

While the evidence from the Siege of Paris may still be up for debate, the same cannot be said for two less famous sieges from the next century. The first of these is the Siege of Senlis in 949. This was a joint French-German operation consisting of Conrad the Red, Duke of Lorraine (r.944–53), serving on behalf of the German King and future emperor Otto II (r.936–73, as emperor from 962), and King Louis IV of West Francia (r.936–54). Their target was the lands of Hugh the Great, Duke of the Franks and Count of Paris (c.898–956), father of Hugh Capet, the future King of France (r.987–96). The Franco-German army, mostly recruited locally and consisting of many Belgians, first burned the city's suburbs and set about trying to take it by force. The contemporary historian Richer of Rheims (fl. tenth century) wrote that the king's soldiers suffered extremely under the crossbow shooting of the town militia and were eventually forced to withdraw, although Richer stressed that it was not only on account of the crossbows that they were forced to retreat. Even when forced to share the credit this was still a clear reference to the use of the crossbow and a clear victory for it over the besieging forces.[21]

The Siege of Verdun in 985 was the result of another incursion into France by a duke of Lorraine, this time by Theodoric I, Duke of Upper Lorraine (c.965–c.1027). The old city of Verdun still had its third-century Roman walls, but in the intervening centuries a large suburban community had spread beyond the old city. Unlike at Senlis, Theodoric successfully captured Verdun and tried to bolster its fortifications against an inevitable French counter-attack. Verdun's position on the river Meuse made it a valuable target, especially for the duke as control of the city would make it easier for him to defend his duchy. King Lothair (r.954–86) assembled an army to drive Theodoric out, and when he reached Verdun, he built a fortified camp from which he could execute the siege. What followed was a form of mutual siege warfare, with both armies assembling massive siege engines and attempting to expel the other from their fortified position. Siege towers, *ballistae* and crossbows were all used in this bloody engagement. Eventually Lothair was successful, and Theodoric was forced to surrender Verdun to the king.[22]

While Verdun is arguably more important for the use of massed siege machinery and the sophisticated siege techniques it showcased, it also contained clear evidence for the use of the crossbow. Just like at Senlis, Richer left an account of Verdun. While he wrote with a clear pro-Lothair perspective, his level of detail is invaluable to modern historians. In his description of the types of weaponry used by the French against incursions from Lorraine he specifically mentioned '*arcobalistae*', another spelling of the term for crossbows used in Vegetius.[23]

In the tenth and eleventh centuries the crossbow went from a relatively obscure weapon, at least in the available sources, to one of the most widely-used ranged weapons in medieval warfare. It is difficult to determine what caused this shift, but it was probably a combination of two factors. The first is that there were just more crossbows in Europe after the year 1000. Warfare and weaponry got more sophisticated over the course of the Middle Ages and rulers were able to assemble better-trained and better-equipped armies in 1300 than they could 500 years earlier. The second factor is that more sources were written and have survived in the later periods, which gives more opportunities for the role crossbows played in warfare to be recorded by contemporary chroniclers. The more evidence historians have, the easier it is to reconstruct the warfare of the period. It is possible that there were more crossbows in the Early Middle Ages than the available evidence suggests, which would mean there was less of a difference between the time periods, but that will remain speculation unless we discover new evidence for the earlier period.

Chapter 5

The Crossbow in the High Middle Ages

In February 1119, the Norman noble Eustace de Breteuil (before 1110–36) petitioned King Henry I (r.1100–35), who was also the Duke of Normandy, for the rights to the castle of Ivry. Eustace made implied threats that he would join an ongoing rebellion against the king if he were not appeased, but Henry was unwilling to give him the castle. Instead, King Henry arranged a hostage swap between Eustace and the current possessor of the castle, Ralph-Harenc.[1]

Eustace received Ralph-Harenc's son and handed over his two daughters in exchange. Eustace, possibly under the influence of Amaury de Montfort – one of the nobles in rebellion against Henry I – blinded Ralph-Harenc's son and sent him back to his father. The father, rightfully appalled and enraged by this mistreatment, petitioned the king to allow him to mete out his own punishment to Eustace's daughters. He asked the king's permission because Henry I was their grandfather. Eustace was married to one of Henry's illegitimate daughters, Juliane de Fontevrault (1090–after 1136), so Ralph-Harenc wanted to be sure he would not displease the king. Despite their blood relationship, Henry I granted permission and Ralph-Harenc had his two hostages blinded and their noses cut off – a truly horrifying punishment for two innocent young women. They were then returned to their parents who were understandably upset.

In the wake of these events Eustace closed his castles to the king as an act of rebellion and sent his wife, Juliane, along with a small force to defend the town of Breteuil. Henry I in response moved to attack Breteuil where the people of the town opened the gates to their duke. In response, Juliane sealed herself into the town's citadel and entered negotiations with her father.

The chronicler Orderic Vitalis, writing less than 25 years after the events, gave the following description of her 'negotiations':

> . . . she treacherously sought a conference with her father, and the king who suspected no such fraudulent design in a woman, giving her the meeting, his unhappy daughter attempted his life. Drawing a crossbow she launched a bolt at him, but through God's protection he escaped unharmed. Thereupon the king ordered the drawbridge of the castle to be broken down, so that no one could either enter or come out.[2]

Orderic provided no detail of the context of the meeting or who else was in attendance. Given that Juliane had hidden the crossbow from sight and successfully escaped after failing to kill her father, the most likely explanation is that she met him from atop the battlements of her citadel while he stood below. The detail about the destruction of the drawbridge could indicate that they met in front of the citadel and Juliane successfully fled back into it, but that still would not explain how she concealed the crossbow.

After her failed assassination attempt Juliane was forced to flee. Orderic wrote that while lowering herself from the walls of the citadel she fell into the ice-cold moat but she still managed to escape successfully and was reunited with her husband. After having their lands revoked by the king due to their rebellion, Eustace and Juliane were eventually pardoned and allowed back into his good graces in the autumn of that year. Eustace died in 1136, the year after Henry I, and Juliane lived out the rest of her life in a nunnery.[3]

In some ways this story is typical of the antagonism found among members of the medieval nobility. Rebellion and reconciliation were common features of medieval politics, and it is no surprise that this extended to the king's family – both 'legitimate' and otherwise. However, in other ways this is a rather extreme example. The horrifying disfiguring of hostages represents an obscene violation of the courtesy usually expected of members of the elite. Orderic Vitalis did not even give a reason for Eustace's blinding of Ralph-Harenc's son, making the action seem all the more irrational. The most extreme detail, however, is probably Juliane's attempted assassination of her own father with a crossbow. Attempted assassination of a medieval king was fairly rare, very rare for the would-be assassin to be the king's daughter and rarest of all for that to have been carried out with a crossbow. The preferred weapon of the medieval assassin was usually the knife or sword, something up close and personal which made death or maiming of the target more likely, even if the assassin had far less chance of escaping.

Sadly, the short length of the account makes it impossible to dig any deeper into the interesting character of Juliane. It would be interesting to know what kind of previous experience she had with crossbow shooting. Was she an experienced archer or just moved by desperation to try and take whatever chance she could? What kind of crossbow did she use? Why did Henry I forgive her so readily after she tried to kill him? Perhaps if Juliane had succeeded then more medieval authors would have written about the incident and provided more detail, such as with the many accounts of the death of Richard I, but since he survived the story has been relegated to relative obscurity.

Crossbows and the Crusades

No medieval conflict lives on as strongly in the modern imagination as the Crusades. The full scope of the Crusading movement extended across all of Europe from the Baltic to Iberia, but the most famous Crusading campaigns were those in Syria and the Middle East, especially between the years 1095 and 1192. The Crusader States established after the successes of the First Crusade had one of the highest rates of continuous warfare of anywhere in the Middle Ages, which makes them an ideal place to study the use of the crossbow in high medieval warfare.

Before the use of the crossbow can be examined, however, it is necessary to establish where and when crossbows were used. As previous chapters have shown, this is easier said than done given the problems interpreting medieval terminology. For the records of the First Crusade the central problem revolves around the use of the term *ballista* and whether authors meant crossbows or torsion-powered siege weapons, an issue that will be familiar from the previous chapter. Colm Flynn has made a thorough and convincing argument showing that in most but not all cases authors writing about the First and Second Crusades used *ballista* to mean a torsion-powered siege weapon. However, we know that crossbows were used during the First Crusade (see for example the description provided by Anna Komnene discussed at the start of Chapter 1), so how can we tell the crossbows from the siege engines?[4]

Some medieval authors used alternative terms that could be interpreted as crossbow instead of *ballista*. For example, Albert of Aachen (fl. 1100) wrote an influential account of the First Crusade drawing on interviews with participants upon their return to Europe. Albert described the use of *arcus Balearis* in several conflicts, for example at the Siege of Sidon in 1110. This has generally been interpreted as meaning crossbows, especially because Albert nearly always used *ballista* to refer to the siege weapon. This interpretation of *arcus Balearis* is not without its problems: it literally translates to Balearic Bow, i.e. a bow from the Balearic Islands off the coast of Spain. Classical writers made many references to the famous slingers of the Balearics, and it is possible that Albert was making an elaborate classical reference in his use of the term. It seems more likely that Albert meant crossbows than slings, but the interpretation of this term is fraught, and historians must show some hesitancy in interpreting it as meaning crossbows.[5]

There are fewer challenges with interpreting terminology in the later Crusade chronicles. By the time those were written a clearer terminology for crossbows had emerged and when combined with the contexts used by those authors historians can be more confident in declaring when an author meant crossbows and when they did not. Still, interpreting medieval chronicles is not an exact science and historians should avoid convincing themselves that they know exactly what an author from 800 years ago thought.

The full story of the Crusades extends well beyond the scope of this book, so instead two case studies will be covered below. First, a discussion of the crossbow in the First Crusade, looking particularly at the terminology used by medieval authors and the problems with identifying the roles the crossbow played in the battles and sieges that took place between 1095 and 1099, and second an examination of how the crossbow was used during the famous

Third Crusade. The three main engagements of that conflict will be examined in detail to show how the crossbow played an integral role in the tactics of Crusading warfare.

The First Crusade (1095–1099)

At the Council of Clermont in 1095, Pope Urban II (r.1088–99) called for the warriors of Christendom to travel to the east to retake the holy city of Jerusalem from Islam. That is one version of his speech anyway – the original text of what he said at Clermont does not survive and instead historians have had to piece it together from later accounts, many of which differ substantially on the details. Many warriors took the cross and agreed to go east to Constantinople to assist the Byzantine Emperor Alexios I against his Islamic opponents before, hopefully, moving on to reclaim Jerusalem for Christendom. For a variety of reasons, no monarchs participated in the First Crusade, but the list of noble princes included the brothers of two kings along with an impressive number of dukes and counts from across Western Europe.

The First Crusade was a campaign defined by major sieges. It was not without its battles, most notably at Dorylaeum in 1097 and at Ascalon in 1099, but it was the sieges that put the campaign in greatest danger and delivered its greatest victories. Three sieges in particular stand out: Nicaea, Antioch and Jerusalem. It is through these sieges that the role of the crossbow in the First Crusade can best be examined. However, it is during these sieges that the terminology used by the chroniclers can be at its most challenging.

The Siege of Nicaea

The Siege of Nicaea in May 1097 was the Crusaders' first real test in the East. The Byzantine army supported them in laying siege to the Seljuk-controlled city. The Crusaders took an aggressive approach, attacking the walls with siege weapons and beginning the process of undermining the walls almost immediately. On 21 May, Kilij Arslan, the ruler of the Seljuk Turks, sent an army to try and relieve the siege but it was driven off after a pitched battle. Nicaea is located on the shores of Lake Askania (Lake İznik in modern Turkey) which prevented the Crusaders from fully blockading it. The Byzantines then dragged ships overland to the lake and by 17 June they had entirely surrounded the city. Now trapped and with their relief army having already been driven off, the Turkish forces surrendered to Emperor Alexios on 19 June, and he took the city without sacking it. The Crusaders were unhappy since they had expected spoils from looting the city. Relations between the First Crusade and the Byzantine emperor would only get worse.[6]

In a letter home, the Crusader Stephen of Blois (c.1045–1102) referred to the use of *ballista* as part of the battle against Kilij Arslan's forces outside the walls of Nicaea. Only two of Stephen's letters survive which makes it difficult to identify a pattern in how he used military terms. In the context of a battle against relief forces it seems likely that he meant crossbows as any siege weapons would have already been in use attacking Nicaea.[7] We do know that siege *ballistae* were used at Nicaea because Raymond of Aguilers (fl. eleventh century), another eyewitness participant who wrote an account of the Crusade, and Albert of Aachen both mention them. So, we have multiple accounts using *ballista* to mean different weapons, which is why it is so challenging for historians to decipher exactly what was happening.[8]

The anonymous author of the *Gesta Francorum*, a highly influential eyewitness account of the First Crusade, mentioned the use of *ballistae* to protect the actions of the siege engineers in their attempts to undermine the city walls. Undermining was an extremely dangerous activity where engineers would dig under the walls of a fortification, sometimes through tunnels but more often by approaching the walls and digging at their base. They would build scaffolding to support the walls as they removed the earth from beneath them and then set fire to the scaffolding, or in later centuries detonate payloads of gunpowder, to collapse the ground underneath the walls and bring a section tumbling down. This was extremely dangerous work as the defending soldiers would try everything they could to attack and kill the engineers. The besieging forces would also often try laying down suppressing fire to keep the defenders off the ramparts and disrupt their ability to attack the engineers. The *Gesta Francorum* usually used the term *ballista* to mean crossbows, and at Nicaea this is the likely interpretation since they would have been better suited to the constant barrage of missiles required to protect the engineers.[9]

The Siege of Nicaea shows the complexity in interpreting First Crusade sources, especially since both siege *ballistae* and crossbows called *ballistae* were present. It also shows two of the most common uses for crossbows in offensive siege warfare: driving off an attack by an enemy relief army and protecting the siege engineers from attack.

The Siege of Antioch

The Siege of Antioch was arguably the most significant single event in the First Crusade: the moment when the Crusaders came closest to being entirely wiped out, lost the support of the Byzantine Empire, and emerged victorious over an enormous Turkish army. The story of the siege contains betrayals, miraculous discoveries, theological debates and several near-death experiences. Unfortunately, it would take far too long to describe every detail, so instead we will have to focus on how the crossbow contributed to the Crusaders' survival.

The abridged version of the siege is that the Crusaders arrived outside the ancient city of Antioch on 21 October 1097 having travelled a great distance since Nicaea and won a victory over the Seljuk Turks at the Battle of Dorylaeum.

The siege of Antioch lasted from October 1097 until June 1098, during which time the two sides engaged in countless small skirmishes and battles. Finally, on 2 June Bohemond of Taranto (c.1054–1111), a Norman Crusader from southern Italy, succeeded in taking the city by treachery. The Crusaders entered Antioch just in time as a massive army lead by Kerbogha (fl. eleventh century), the Atabeg of Mosul, arrived outside the city on 5 June and immediately put the Crusaders under siege. The newly-besieged defenders were perilously low on supplies after almost a year of siege warfare. On 28 June the Crusaders, their morale bolstered by the possible discovery of the holy lance that had pierced Jesus' side when he was on the Cross, sallied out of Antioch against Kerbogha's much larger force and, much to Kerbogha's surprise, drove the Muslim leader off – his army disintegrating in the face of the Crusaders' charge. The exact reasons for Kerbogha's failure have been extensively debated, but blame can be placed in part on underestimating the determination and desperation of the Crusaders who faced total annihilation if they did not win, as well as the nature of Kerbogha's army – many of the rulers he had brought with him on campaign were happier to watch Kerbogha fail than they were to risk their lives in battle against the Crusaders over a city that they would not even get a share of if he was victorious.[10]

In his account of the First Crusade, the French abbot Baldric of Dol (c.1050–1130) mentioned the use of *ballistae* siege weapons at the siege of Antioch, and, since his account was based on eyewitnesses he interviewed, this was probably accurate. To add confusion to Baldric's terminology, in his later description of the siege of M'arrat-an-Numan, infamous for a supposed incident of Crusader cannibalism, he seems to have used *ballista* to mean crossbows when he described Crusaders holding them, a frustrating inconsistency in terminology that is by no means unique to Baldric.[11]

The chronicler William of Tyre (c.1130–86), who was born in the Crusader States and wrote his history almost a century after the First Crusade, mentioned *ballista* in his account of the Siege of Antioch. He described the Crusaders using *ballista* as they sallied out to battle Kerbogha, which suggests they were crossbows as the dragging-out of siege weaponry would have been highly impractical. He did not provide detail about how the crossbows were deployed, but it is tempting to think that they were pushed out first to cover the full deployment of the Crusading forces into battle formation – possibly providing covering fire to keep the Muslim soldiers at bay while the army emerged from Antioch.[12]

As with Nicaea it seems that both siege *ballistae* and crossbows were used at Antioch. Antioch is a great example of the fluid nature of siege warfare, where an army could go from being outside the walls to being trapped inside them in only a matter of weeks and the deployment of their weapons would have had to shift with the new challenges each twist and turn in the siege presented.

1. Cross section of composite crossbow, fifteenth century, Metropolitan Museum of Art no. X.800. (Designated CC0 Public Domain)

2. Crossbow nut, pre-1272, Metropolitan Museum of Art no. 28.99.28. (Designated CC0 Public Domain)

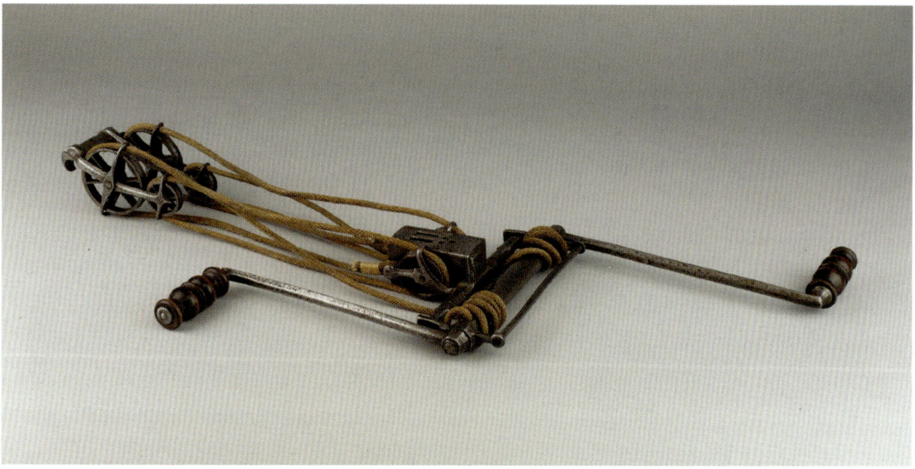

3. Windlass, c.1450–1600, Art Institute Chicago no. 1982.3087. (Designated CC0 Public Domain)

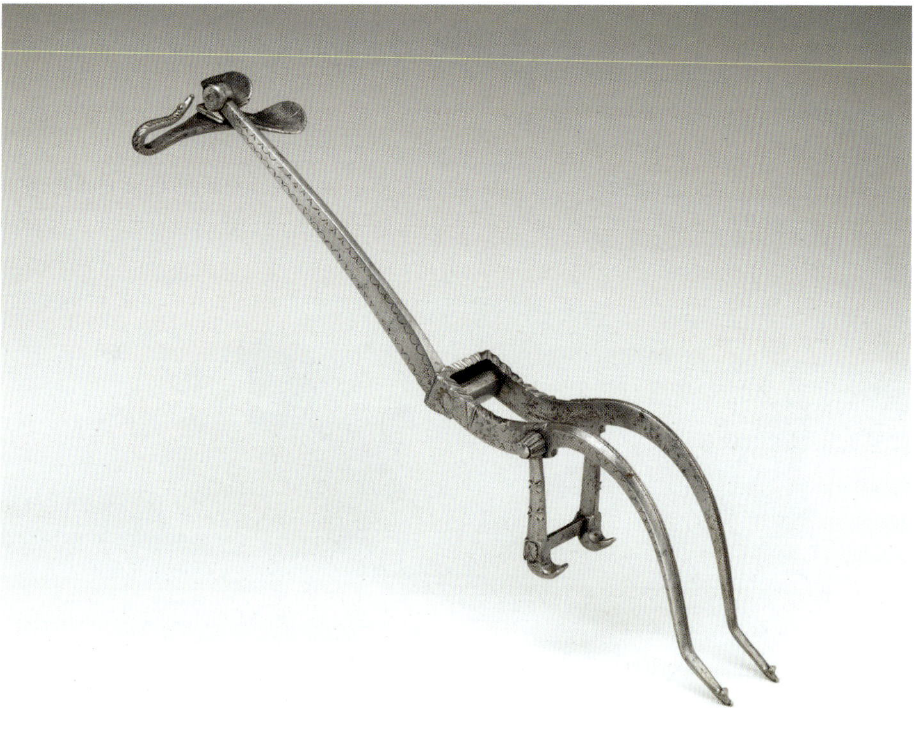

4. Goats-foot lever, late fifteenth or first half of the sixteenth century, Metropolitan Museum of Art no. 14.25.1609. (Designated CC0 Public Domain)

5. Crossbow bolts for war or hunting, probably late fifteenth or early sixteenth century, Metropolitan Museum of Art, nos. 14.25.1603, 14.25.1600, 14.251597, 14.25.1595, and 14.25.159.9. (Designated CC0 Public Domain)

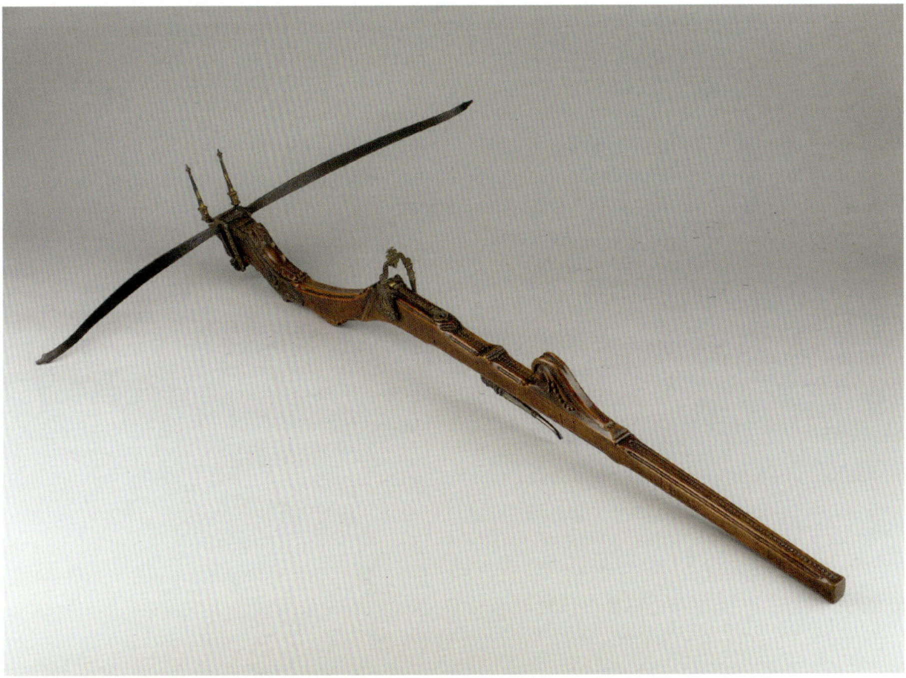

6. French bullet crossbow, c.1570–1600, Art Institute Chicago no. 1982.3090. (Designated CC0 Public Domain)

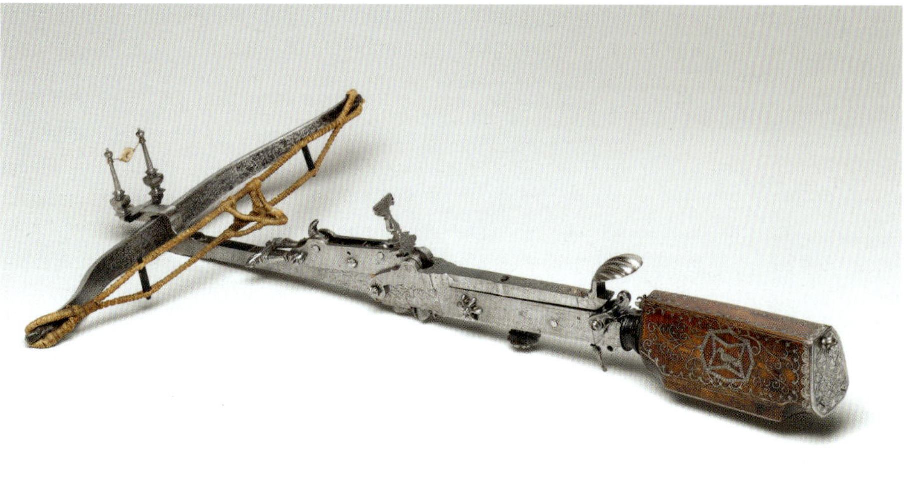

7. German bullet crossbow, c.1600–50, Metropolitan Museum of Art, no. 14.25.1588. (Designated CC0 Public Domain)

8. *Balestrino*, probably seventeenth or eighteenth century, Metropolitan Museum of Art no. 54.46.4. (Designated CC0 Public Domain)

9. Bullet and bolt crossbow combined with a wheel-lock gun, c.1570–1600, Metropolitan Museum of Art no. 14.25.1573. (Designated CC0 Public Domain)

10. Beatus of Liébana's *Commentary on the Apocalypse*, Cathedral Library Burgo de Osma, Cod. 1, fol. 85v. (Public Domain Mark, Wikimedia Commons)

11. Hunting scene, first half twelfth century (possibly 1129–34), Metropolitan Museum of Art, New York City. (Designated CC0 Public Domain)

12. *Morgan Picture Bible*, MS M.638 10v, Morgan Library, New York City. (Public Domain Mark, Wikimedia Commons)

3. *Romance of Alexander*, Bodleian Ms. 264, fol. 81v. (Public Domain Mark, Wikimedia Commons)

14. Dunois Master, Life and Death from the *Coëtivy Hours*, Chester Beatty Library W 082 f.317. (© Chester Beatty Library, used with permission)

15. Battle of Crécy in Froissart's *Chronicles*, Bibliothèque nationale de France, Fr 2643, f.165v. (Public Domain Mark, Wikimedia Commons)

16. The Battle of La Higueruela by Fabrizio Castello, Orazio Cambiasi and Lazzaro Tavarone in the Gallery of Battles at the Royal Monastery of San Lorenzo de El Escorial. (Public Domain Mark, Wikimedia Commons)

17. Meister der Heiligen Sippe d. J, Martyrdom of St Sebastian, c.1480, Bildarchiv & Wallraf-Richartz Museum, Cologne. (Public Domain Mark, Wikimedia Commons)

18. Piero de Pollaiuolo, Martyrdom of St Sebastian, c.1475, National Gallery, London. (Public Domain Mark, Wikimedia Commons)

23. Crossbow of Matthias Corvinus, King of Hungary (r.1458–90), 1489, Metropolitan Museum of Art no. 25.42. (Designated CC0 Public Domain)

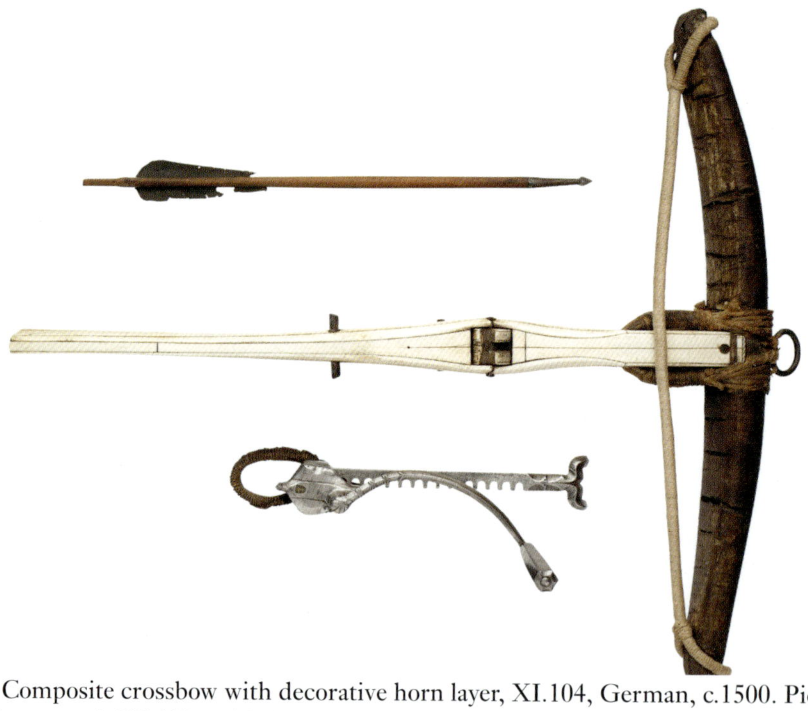

24. Composite crossbow with decorative horn layer, XI.104, German, c.1500. Pictured with *cranequin* XI.105 and bolt XI.121, Royal Armouries, Leeds (© Royal Armouries, used with permission).

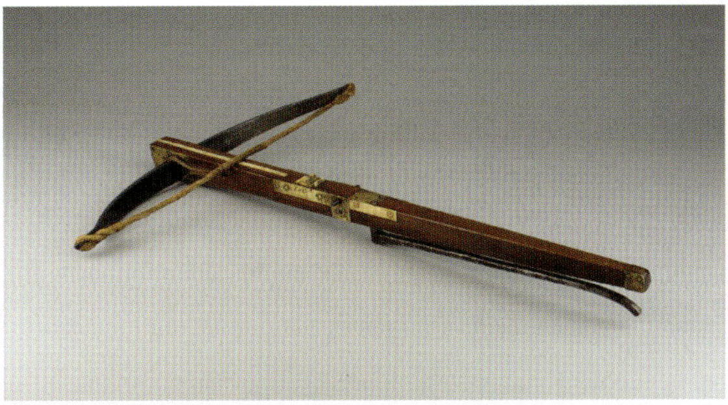

25. Spanish crossbow, c.1530–1610, Art Institute Chicago no. 1982.3073. (Designated CC0 Public Domain)

26. Han Dynasty crossbow trigger, Henan Provincial Museum, Zhengzhou. (Creative Commons CC0 1.0 Universal Public Domain Dedication)

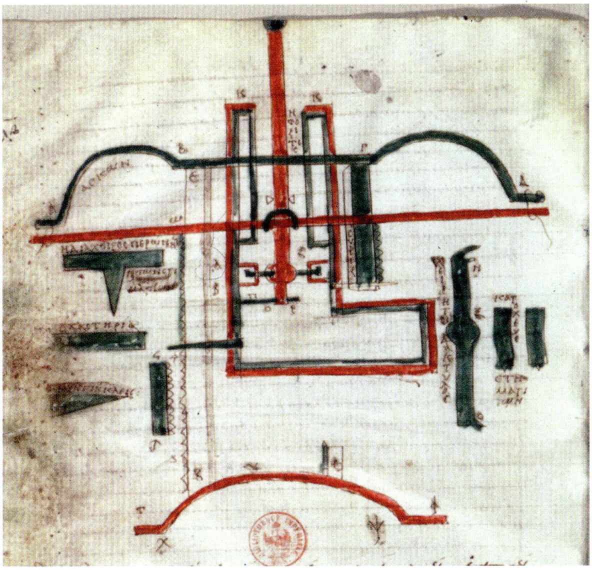

27. Diagram of a *gastraphetes*, Heron of Alexandria: 'Belopoeica', *Copy M: Codex Parisinus inter supplementa Graeca 607* f.47v, Bibliothèque nationale, Paris. (Public Domain Mark, Wikimedia Commons)

28. Battle of Sluys in Jean Frois-sart, *Chronicles*, Bibliothèque nationale de France, MS Fr. 2643, Folio 72r. (Public Domain Mark, Wikimedia Commons)

29. Malta before the Great Siege, Dimostrazione di Tutte le batterie by Mattia Perez d'Aleccio, 1581, Grand Hall of the Palace of the Grand Masters, Valletta. (Public Domain Mark, Wikimedia Commons)

30. Hunting Scene from Gaston Phoebus, *Livre de Chasse*, Paris, Bibliothèque nationale de France MS fr. 616, fol. 115r. (© BnF, us with permission)

The Siege of Jerusalem

The culminating moment of the First Crusade was the taking of Jerusalem in 1099. The crusade had temporarily lost its momentum after Antioch and only a portion of the participants made it to Jerusalem. Bohemond stayed in Antioch, claiming the city for himself, while the future King Baldwin I (r.1100–18) was in Edessa, the city where he had delayed Kerbogha's army long enough for Bohemond to take Antioch.

The remnants of the Crusading army, led primarily by Count Raymond of Toulouse (1041–1105) and Duke Godfrey of Bouillon (c.1060–1100), arrived outside Jerusalem on 7 June 1099. The Crusader force was badly undersupplied and the environment around Jerusalem lacked much wood with which to build siege weapons. The Crusaders were saved on 17 June with the arrival of a small Genoese fleet in the nearby port of Jaffa. The Genoese became trapped in the port and decided to dismantle their ships and bring them overland to the Crusaders, providing enough wood for the construction of two siege towers. On 14 July, the Crusaders attacked the city. One siege tower, commanded by Duke Godfrey, successfully reached the city's walls and from there Crusaders poured in, massacring enemy soldiers and civilians alike.[13]

The siege tower was a staple of Western European siege warfare. It was a tower on wheels that could be slowly pushed up against the walls of the city. The tower had multiple functions. Once it was against the wall attackers could climb up inside it and emerge out onto the walls. At the same time, the top of the tower was usually filled with archers and crossbowmen who could shoot at defenders on the battlements from higher up – a task that was very important to keeping the tower intact as defenders would try everything to destroy it, usually by trying to set it on fire with either flaming arrows or naptha, an oil-based flammable substance that was thrown in jars and then set alight. Lastly, the base of the tower could protect siege engineers who would work to undermine the walls. At Jerusalem, the deciding fact was Godfrey of Bouillon's tower reaching the walls and a force of Crusaders, including Godfrey, entering the city and securing the gates to let in the rest of the army. No accounts of the siege single out the crossbowmen for their role in the siege tower, but we know from other sieges, in both the Crusader States and Europe, that they played an integral role in the successful delivery of a siege tower to a city's walls.[14]

The Third Crusade (1189–1192)

The Third Crusade can easily claim to be the most iconic of the many Crusader campaigns. Called as a response to the capture of Jerusalem by Saladin in 1187, it saw all of the major monarchs of Europe united in an effort to resist the sultan's conquests and, hopefully, retake the holy city. Emperor Frederick I (r.1155–90) died while traveling east, never making it to the Holy Land and causing many German Crusaders to return home. King Philip II of France and

King Richard I of England were more successful and both made substantial contributions to the Crusade, although Richard's legend would eventually overshadow the French king's role. Both monarchs were central to the first, most famous and most successful engagement of the Crusade: the Siege of Acre. After the successful siege Philip returned to France while Richard continued his campaigning in the Holy Land. Richard failed to retake Jerusalem but successfully recaptured the port of Jaffa and defeated Saladin's armies on two occasions.

Steve Tibble has argued that over the course of the twelfth century warfare in the Crusader States took on a distinct form, driven in a large part by the requirements of fighting against Turkish nomadic horse archers. Tibble also stressed that the new requirements of Crusading warfare created a greater emphasis on the crossbow and other ranged weaponry. A close study of the three major conflicts of the Third Crusade – the Siege of Acre and the Battles of Arsuf and Jaffa – provides the ideal venue for examining how Crusading warfare had developed during this period and the crossbow's prominent role therein.[15]

The Siege of Acre

The Siege of Acre was one of the most impressive events of the entire Middle Ages, a massive siege that lasted more than two years and saw three Christian monarchs opposing one of Islam's greatest sultans. The Christian forces included participants from all over Europe, most famously the kings of England and France but also Germans, Danes, Italians, Hungarians and many more. The scale and length of the conflict makes it ideal for studying the many techniques involved in medieval siege warfare – nearly every piece of technology and strategy available to a twelfth-century army was used at Acre. At the same time, its massive scale and length make it nearly unique – no other conflict was quite like the Siege of Acre. In terms of the crossbow, it was used in the same ways already discussed in the First Crusade but on a grander and more mature scale than in those earlier conflicts.

The siege began in late August 1189 when Guy of Lusignan, King of Jerusalem (c.1150–92, r.1189–92), without a capital city since 1187 and having been locked out of Tyre by Conrad of Montferrat (1146–92), decided to lead a force of only a few thousand men, including members of the Knights Templar and Knights Hospitaller, from outside Tyre towards Acre. Acre was one of the most valuable ports in the Middle East, an ancient city located on a peninsula jutting into the sea with massive walls and a well-protected harbour. Guy's army reached the city on 28 August having been left largely unopposed by Saladin who was besieging another city at the time. Guy's army, recently reinforced by the arrival of some Pisans, opened the siege with an assault on 31 August but was repulsed. His first attack a failure, Guy decided to settle in for a long siege.

Saladin eventually moved his army to a position of strength near Acre from which he could reinforce the garrison and threaten Guy's army. Despite his numerical strength Saladin elected not to attack Guy, which was to prove a mistake for the sultan as the Crusading army would only grow with time despite suffering significant losses throughout the siege. The first wave of reinforcements consisted mostly of Danes and Frisians who arrived in September, and for the following two years from spring until autumn a steady stream of Crusaders and supplies would arrive from Europe. The most significant, if much delayed, arrivals were Philip II of France in April 1191 and Richard I of England in June of that same year. This constant supply of new manpower meant that despite Saladin's numerous attacks on the camp and the high rate of attrition among the Crusaders, they were never short of soldiers with which to prosecute the siege.

From the end of October 1189, the Crusading army set about fortifying the camp with ditches and palisades. This created a siege within a siege situation where the Crusaders were simultaneously besieged by Saladin's army and besieging Acre. The Crusaders constructed numerous catapults and other siege weapons which they used to attack Acre and were in turn attacked by catapults from within Acre. They also had access to their own stretch of coast where ships from Europe could land. The siege featured an important maritime component as the Crusaders' navies were necessary for both their own supply and for cutting off the supply of Acre. Many major naval engagements occurred off the coast of Acre, usually ending in the Crusaders' favour, making Acre really the full package in terms of medieval warfare.[16]

Throughout the siege both sides engaged in multiple skirmishes and battles as Saladin tried to crush the Crusaders' camp and the Crusaders kept putting pressure on Acre. John Hosler has estimated that the two sides fought eight major battles as part of the Siege of Acre, a number that would be impressive for any medieval military campaign, as well as twenty-six skirmishes and many more sorties, undermining attempts and naval actions. The Crusaders even took the offensive at times, launching their own attacks against Saladin's camp as well as against Acre.[17]

As was common in any lengthy siege disease quickly became a greater threat than enemy soldiers, and thousands of Christians and Muslims alike died of illnesses throughout the two-year siege. Things were at their worst during the winter when supplies became hard to find due to the difficulty of sailing the eastern Mediterranean in that season. Both sides suffered significant privations from lack of food. Saladin dispersed large portions of his army to try and limit the suffering and expense of keeping it in the field for so long, but the Crusaders were forced to endure starvation, disease and miserable weather for months on end. The coming of the spring brought new troops, supplies and money for both sides in the conflict.[18]

On 12 July 1191, after multiple breaches had been made in its walls, the garrison of Acre surrendered on the terms that Saladin would pay the kings 200,000 golden dinars, return the piece of the True Cross lost at Hattin in

1187, and release all of the Frankish prisoners he held. The Muslim garrison was to be held hostage until Saladin met these terms. The surrender came to a terrible end as Richard ordered the massacre of most of the Muslim garrison when he felt that Saladin was not being forthcoming with his end of the agreement, a brutal and highly controversial decision. Saladin massacred his Frankish prisoners in response and abandoned the rest of the terms of the surrender.[19]

Throughout the siege the crossbow made a significant contribution to hostilities on both sides. To describe every time the crossbow was mentioned or likely used during the two years of the siege would be a nigh endless task, instead we will focus on a handful of representative cases that best illustrate the different ways the crossbow contributed to the conflict before concluding with a few interesting anecdotes involving the crossbow at Acre.

As mentioned earlier, the siege tower was a staple of Crusader siege tactics, and the crossbow often played a key support role as part of the process of successfully delivering a siege tower to the walls of the city. The Crusaders built three towers in the spring of 1190. Richard of Templo (fl. 1220s), wrote his *Itinerarium Peregrinorum et Gesta Regis Ricardi* in the early thirteenth century but based it on other eyewitness accounts, described how the top levels of the towers were filled with archers and crossbowmen.[20] Unfortunately for the Crusaders, while they managed to get the siege towers to the city's walls in early May, the defenders successfully set fire to and destroyed them before the walls could be taken. The most interesting example of a siege tower at Acre was when the Crusaders constructed a siege tower on a ship and used it to try and capture the Tower of Flies, one of the two towers that flanked Acre's harbour. A fleet of Christian galleys sailed out with the tower-ship to defend it from the Muslim ships. The siege tower was manned by Pisans and filled with missile troops, probably including many crossbowmen, who were supposed to distract the defenders so that the rest of the soldiers could climb into the Tower of Flies from a position of comparable height. The plan almost worked as the siege tower was anchored next to the Tower of Flies and several Crusaders climbed onto the tower as an archery duel raged around them. However, the Muslim defenders managed to destroy the tower-ship using Greek Fire and the operation ended in failure.[21]

At Nicaea, the Crusaders had used missile barrages to protect the siege engineers in their attempts to undermine the city's walls. The same thing was common at Acre but on a much larger scale. After his arrival in April, King Philip II ordered that the archers and crossbowmen should launch a near-constant barrage of missiles into Acre night and day. He also ordered the construction of new catapults and once they were complete, they added to the steady stream of projectiles flying from the Crusader camp into the city. Ambroise (fl. 1190s), a Norman who travelled east with Richard I and wrote a verse history of the

Third Crusade in French, wrote that Philip had a protective shelter built for himself and he would often sit under it and shoot his crossbow at the defenders of Acre.[22]

Crossbows played an essential role in defending the Crusader camp from attacks by Acre's garrison and Saladin's army – often at the same time – as well as supporting Crusader counter-offensives. This happened countless times over the course of the siege, but one notable example of both Crusader defence and counter-attack occurred in autumn 1189. Saladin attacked the Crusaders on 16 September, but his soldiers faltered in the face of a withering barrage of crossbow bolts and a wall of Crusader spearmen. The Crusaders launched a counter-attack on 4 October, with the archers marching at the front followed by the foot soldiers and the elite cavalry in the back waiting for the opportunity to unleash their charge. The battle started at nine in the morning and raged all day but despite initial successes the Crusaders were eventually repulsed with both sides suffering heavy losses. The crossbow showed its capacity as both an offensive and defensive weapon on those days.[23]

While this section has mostly emphasized the use of the crossbow by the Crusading armies, it was equally common in Muslim armies at the time. Unfortunately, the Muslim accounts of the Crusade do not provide the same level of detail on the role of the crossbow in contemporary warfare. It is clear from what evidence we do have that they used the crossbow extensively, and they used them in conjunction with composite bows. In sieges they would use crossbows to protect their miners and siege engineers just like the Crusaders – William of Tyre recorded Saladin doing exactly that when he besieged Beirut in 1182.

The Muslim author al-Tarsusi wrote a manual on warfare in the late twelfth century and dedicated his work to Saladin. In his book is possibly the earliest reference to a windlass, showing that Muslim armies were not only using the crossbow but potentially pushing the technology to new heights. There is also a reference to what was probably a crossbow with a windlass at the Siege of Acre. Ambroise recorded that a Muslim defender shot his *arbalast a tur* at a Christian man-at-arms and struck him in the chest piercing his hauberk, but the man-at-arms was saved because the bolt hit a letter hanging from his neck on which was inscribed the names of God. The story is told in a section of Ambroise's account that predates his arrival at Acre, so he was repeating this fanciful series of events second-hand. However, even if the actual event probably did not happen, the weapon chosen for it probably did reflect a real missile weapon that many Muslim defenders were using against the Crusaders. *Arbalest a tur* was usually the French equivalent of the *ballista ad turnum* discussed in Chapter 2 and probably refers to a windlass-spanned crossbow, although it could possibly refer to a wall-mounted great crossbow. While both pieces of evidence suggest a Muslim origin for the windlass crossbow, and that very well may be the case, the fact that the Norman Ambroise already had a term to describe the weapon suggests that it was at least known in the West.[24]

The various authors who were witness to or who wrote contemporary accounts of the Third Crusade provide many other interesting anecdotes about the crossbow at the Siege of Acre. These do not necessarily fit into any broader pattern of the military role of the crossbow, but they are enlightening in their own way. Ambroise wrote about how Richard, while still recovering from an illness soon after reaching Acre, took to sitting under a protective canopy and shooting his crossbow at the city's defenders in much the same way as Philip had. Ambroise also wrote of how one Aubery Clément declared one day that he would either enter Acre that day or die a martyr, a praiseworthy declaration in the Norman author's eyes, and died in a glorious assault on the Cursed Tower, one of the many fabulously named towers in old Acre. Soon after a Muslim defender emerged on the walls wearing the armour and insignia of Aubery Clément and as punishment for this transgression Richard shot him in the chest with his crossbow. Ambroise also told a story of how one day a Turkish defender supposedly came on to Acre's battlements with a defiled cross from one of the city's churches and decided he would urinate on it in view of the Crusader armies. He was punished for this insolent behaviour by a Christian archer who shot him in the stomach and bowels with a crossbow.[25]

Throughout Ambroise's anecdotes of individual feats of marksmanship is a theme of the crossbow acting as a tool of God's divine judgment. God saved the man-at-arms from the Muslim's bolt and God punished the Muslims who defiled Christian chivalric reputation or the holy cross itself. This theme of divine punishment and crossbows is possibly at its most explicit when Ambroise wrote about Richard's execution of the garrison. Ambroise was quick to excuse Richard's action and to blame Saladin for forcing the king to execute his prisoners, but he also wrote: 'Thus was vengeance taken for the blows and the crossbow bolts. Thanks be to God the Creator.'[26] Ambroise explicitly linked Richard's execution with divine judgment and punishment for their resistance to Christian attack, but he also singles out the crossbow bolts as particularly warranting revenge.

The Battle of Arsuf

In his book on Crusader armies, Steve Tibble argued for the fighting march as one of the defining tactics of twelfth-century Crusader warfare. A Crusader army on the march was subject to near-constant harassment and attack by highly mobile raiding forces armed primarily with bows and mounted on fast horses. The fighting march was a logical development to protect against these attacks and to effectively deliver the army to the main battle or siege it had been assembled for. The key to a fighting march was having a well-disciplined vanguard and a rearguard that could fight while keeping an orderly march.

At the same time everyone had to be moving at approximately the same speed: if the vanguard marched too far ahead or the rearguard fell behind, they could be isolated and destroyed. Lastly, a core component of missile weapons was key to pushing back the attacking horse archers and limiting their ability to effectively harass the main body of the army. In this final role Tibble noted that the crossbow was particularly popular among Crusader armies as an effective tool for threatening the harassing Turkish cavalry, and in doing so played an essential role in Crusader warfare.[27]

The fighting march played a major role in several Crusader campaigns. The Crusaders first used it in 1099 when the armies of the First Crusade marched out of Jerusalem to face an Egyptian army outside of the city of Ascalon and they would use it again many times over the following two centuries. The fighting march was not unique to the Crusaders: the Byzantines claimed to have invented it and Emperor Alexios I performed one in 1116. The most famous medieval fighting march was led by Richard I in 1191 when he left Acre and marched south along the coast towards the port city of Jaffa.[28]

On 22 August 1191 Richard and a much-reduced Crusading force set out on the march to Jaffa. Many Crusaders preferred to stay in the comfort of Acre, especially after the gruelling siege, but Richard and King Guy managed to assemble a respectable force, including substantial numbers of both Templars and Hospitallers who would take the position of vanguard and rearguard for the march. The more disciplined members of the military orders were ideally suited to the difficult task of maintaining cohesion on the march and were often assigned that role. Archers and crossbowmen were tasked with providing harrying shots to keep Muslim skirmishers back at a safe distance while the army marched. A fighting march was a slow way to travel so the army had to take frequent breaks, often advancing only a handful of miles every day. Richard also had a fleet of ships sailing along the coast that could periodically resupply the army.[29]

The Crusaders were harassed almost constantly as they marched but they also had to fend off regular organized attacks. The first major attacks came on 25 August just outside the city of Haifa and on 30 August, near the city of Caesarea. The Crusaders were able to drive off both attacks with only minimal losses thanks to the fighting march formation. The real test of Richard and Guy's army came on 3 September. The army had just left the Dead River region and emerged onto a large plateau. This time Saladin himself decided to command the attack on Richard's army. He took three divisions and bombarded the Crusaders with a hail of arrows before ordering a general charge. Saladin's forces were unable to drive the Christians back, nor to deliver a fatal blow to their army, but they did achieve at least one minor success: they wounded Richard. Ambroise recorded that Richard was wounded in his side by a *pilet* which could refer to either a crossbow bolt or a light javelin. The wound was minor, and

Richard was able to continue fighting, but it was a close brush with death, and a sign of just how dangerous this campaign could be – had Richard died or even just been incapacitated it is likely the army would have faltered and gone home rather than pushing on to Jaffa.[30]

The main battle of the campaign came on 9 September, just outside the forest of Arsuf only 25 miles from Jaffa. Saladin decided to attack Richard's army again before it reached Jaffa. His army surrounded the Crusaders and showered them with arrows. Richard's army marched through this hail of arrows relying on their armour to protect them in an attempt to reach secure ground near Arsuf. Around the time the Templars in the vanguard reached the position where Richard hoped to set up his camp, the Hospitallers in the rearguard, including a large contingent of crossbowmen, grew tired of enduring the Muslim attack while doing nothing and charged Saladin's lines. Once one portion of his army had broken ranks Richard had to fully commit to the battle or risk destruction, so he had his army turn around and charge Saladin. The Crusaders managed to drive Saladin's army from the field in the ensuing chaos. After his defeat Saladin was not in a position to stop the completion of their march to Jaffa. The city was easily taken since Saladin had torn down its walls in 1190.[31]

Whether Richard intended to confront Saladin at Arsuf, or whether it was mere circumstance and bold decision-making on the part of his rearguard, has been debated back and forth by historians. In Ambroise's version it was not the king's intention to fight that day and while they were victorious the fact that the rearguard engaged too early prevented what could have been an even more glorious victory. A letter Richard wrote several days after the battle also suggests that it was not his ideal outcome. Battles were risky and military wisdom of the time recommended avoiding them unless you were entirely confident that you could win – usually by having overwhelming superiority in terms of numbers and terrain. Nevertheless, the Battle of Arsuf was a resounding success for the tactics of the fighting march and showed how it could deliver good results even when things did not exactly go to plan.[32]

The Battle of Jaffa

On the back of his success at Arsuf, Richard succeeded in recapturing the city of Ascalon as well as a handful of other fortifications in the region, but the campaign stalled before it reached Jerusalem. This was not necessarily the fault of Richard or Guy: laying siege to Jerusalem was a daunting prospect and would require the army marching far inland and exposing its supply lines to significant risk. After twice marching towards Jerusalem before turning around, Richard returned to Acre to consider his options. In response, Saladin launched what would be the final offensive of the Third Crusade.

In July 1192, while Richard was still in Acre, Saladin attempted to seize Jaffa. His army successfully breached the city's walls but were unable to take the citadel. Richard assembled a fleet to sail his army down to the port city, and after a few days of unfavourable winds took off to save Jaffa. Richard performed one of the Middle Ages' few opposed amphibious landings – Ambroise described the king leaping into waist-high water and shooting his crossbow as he stormed the city. After successfully driving off the attackers Richard positioned his army to protect it from Saladin's next attack.[33]

Richard's army was composed of a minority of knights, many of whom had travelled from nearby Caesarea after Richard had relieved the city, supported by a larger force of spearmen and crossbowmen, many of them Pisan and Genoese marines from the fleet that had brought him to Jaffa. Facing a numerically superior Muslim army, Richard had them take up a strong defensive position. He had the spearmen form a shield wall with the tips of their spears protruding between the shields to keep the Muslim cavalry at bay. Behind every two spearmen he placed a pair of crossbowmen. The first crossbowman would shoot while the second was tasked with reloading the weapons and passing them to his companion. This allowed the first crossbowman to constantly have a loaded weapon while both were protected by the shields of the spearmen. Saladin's army charged this formation many times but were driven back each time, suffering casualties at the hands of the crossbowmen as they did. After several thwarted charges Richard switched to the offensive. The crossbowmen led the way, shooting their weapons as they advanced while the cavalry formed up behind them readying their devastating charge. This final push drove Saladin's army from the field and secured Jaffa for the king. The importance of the crossbowmen to this victory was singled out by the contemporary chronicler Ralph of Coggeshall who noted their courage and central role in the king's victory. The Battle of Jaffa shows how the crossbow could be used in both a defensive and offensive manner, and how it was possible to change between the two even in the chaos of battle. It also shows how missile fire could greatly contribute to the shield wall.[34]

Negotiations resumed after Saladin's defeat, and both sides eventually agreed to a three-year truce on 2 September. Saladin recognized the Crusaders' control of the coastal strip between Jaffa and Acre and agreed to let pilgrims visit Jerusalem while Richard had to tear down the walls of Ascalon – a significant concession as he had only just finished rebuilding them. This agreement marked the end of the Third Crusade and Richard returned to Europe soon after, where he would be imprisoned in Austria before eventually returning home in 1194. Saladin barely outlived the crusade, dying on 4 March 1193, having successfully seen the Crusaders off and securing his dynasty for the next century.[35]

The Crossbow in Crusading Warfare

The above examples are hardly the only time the crossbow played a prominent role in Crusading warfare. It is worth considering in brief a few other standout examples to illustrate how central the crossbow was to warfare during this period.

While the Second Crusade has largely been regarded as a failure, having fallen into dissenting factions before failing to take the city of Damascus, it did have one bright spot in the successful capture of Lisbon. Despite happening a continent away, this event is often grouped in with the Crusade because a group of warriors on their way from England to participate in the Second Crusade stopped off in Portugal and assisted in the siege, with most of them returning home after. One of these participants wrote an eyewitness account which he sent back in a letter, now known as *De Expugnatione Lyxbonensi*. In this letter, the author made several references to *ballista* being used as part of the siege, all of which seem to refer to crossbows specifically.[36]

Louis IX of France was particularly fond of the crossbow and brought significant numbers of crossbowmen with him on his two Crusades. Louis created the position of the Grand Master of Crossbowmen, who oversaw engineers and artillerymen as well as crossbows. For the Seventh Crusade he established a crossbow-wielding confraternity at Châteaudun and brought 5,000 crossbowmen with him to Egypt, surpassing the 4,000 crossbowmen who had participated in the Fifth Crusade 30 years earlier. Louis also paid for crossbowmen to stay and assist in the defence of the Holy Land when he returned to Europe.[37]

Banning the Crossbow

At the Second Lateran Council in 1139 the papacy banned the use of the crossbow, or so the story goes. There is some truth to this story, but it is often wrapped in myth and a mistaken understanding of medieval history and the crossbow's role in it. Pope Innocent II (r.1130–43) and the council did issue a decree banning the use of the bow and the crossbow in inter-Christian warfare, that part is true. However, the explanations for why they did so and the impact of this decree are often mistaken. A popular myth is that the crossbow was a threat to Europe's nobility as any common peasant could use a crossbow to kill a noble knight with minimal effort. In reality the reasoning and result of the council were a little more complicated than that.

The Second Lateran Council was the tenth ecumenical council of the Catholic Church, taking place 17 years after the First Lateran Council. It was called by Pope Innocent II with the intention of resolving a papal schism that had arisen after the death of Pope Honorius II (r.1124–30) in 1130. In the wake of Honorius II's death a majority of cardinals and the citizenry of Rome had elected Pietro Pierleoni (died 1138) as pope, but a minority of cardinals selected Innocent instead. This led to a schism in the Church between Innocent II and 'Anacletus II', as Pietro was known. Despite his initial lack of support Innocent II managed to rally significant backing for his claim to being the true pope while in exile in France. He even managed to convince St Bernard of Clairvaux (c.1090–1153), possibly the most influential religious figure at the time, to support him as well as Emperor Lothair III (r.1133–7). Now more politically powerful, Innocent II hoped to crush his rival at the council. However, 'Anacletus II' died in 1138 which ended the schism and left the Second Lateran Council free to pursue other business.

The Second Lateran Council issued thirty canons, many of them reaffirming canons issued by previous councils. Of these, it is the 29th Canon that has drawn the attention of military historians. This states: 'We prohibit under anathema that murderous art of crossbowmen and archers, which is hateful to God, to be employed against Christians and Catholics from now on.'[38] That text was certainly a ban of sorts on both bows and crossbows but with the very important qualifier that it was only for wars between Christians. This council was meeting only 40 years after the conquest of Jerusalem by the First Crusade. One of the Church's motives behind the Crusading movement was to use them to redirect secular violence away from Christendom and towards an external enemy. This context is important for reading the original decree. Innocent II was trying to limit the nobility's capacity to wage war against each other so as to push their fighting to the borders of Europe.

This theme of limiting secular violence can be seen in other canons declared by the council. For example, Canon 12 declared that a truce should be observed between sunset on Wednesday until sunrise on Monday, basically limiting fighting to only three days a week, as well as placing a longer truce in place for major feast events like Advent and Easter. The Second Lateran Council was also interested in limiting other popular activities of the nobility. Canon 14 banned jousting and tournaments, for example. That said, the vast majority of the canons were concerned with the major clerical bugbears of the time: simony, i.e. the buying and selling of clerical offices, and clerical marriage.[39]

In contrast to the eventual success the Church had in limiting both simony and clerical marriage, Second Lateran's impact on the behaviours of the secular population was negligible. Jousting, tournaments, crossbows and warfare on days between Thursday and Sunday continued to be commonplace during the twelfth century and arguably only became more popular later in the Middle Ages. Second Lateran provides an interesting glimpse into the beliefs and policies of the Catholic Church and its clergy at the time, but the Church did not have universal power to dictate the behaviour of the secular nobility. It would be more accurate to view these canons as declarations of how the papacy wanted to see the world run, rather than as enforceable laws.[40]

But what about the idea that the crossbow posed a risk to Europe's elite and the Church was determined to quash that? This argument fundamentally misunderstands several key aspects of both the Second Lateran Council's intent and the crossbow's role in medieval warfare. The Second Lateran Council drew inspiration from the Peace of God movement, which was a clerical movement concerned with stopping knights and other soldiers from attacking and murdering peasants, women and members of the clergy. While the Peace of God would by far have preferred that no killing happen at all, it was a movement concerned with the violence inflicted by the noble elites on the rest of the population, not the other way around. Societal violence in the Middle Ages primarily went one way and while the clerical elite who were at Second Lateran would not have endorsed a peasant uprising – peasants were expected to know their place – they were not attempting to prevent one from happening.[41]

Even beyond the council's goals, it would have been rare for a peasant to get their hands on a crossbow in the first place. Crossbows were expensive weapons used by relatively high-status warriors. They were most famously used by elite mercenary companies and crossbowmen tended to wear heavier armour and command a higher daily wage than your average archer. Evidence of the pay of crossbowmen from twelfth and early thirteenth-century England puts the average wage of a crossbowmen as approximately equivalent to that of a sergeant, well above that of a standard foot soldier, clearly reflecting their elite role in medieval warfare. This continued to be true in the Later Middle Ages as well. For example, in 1451 Edmund Beaufort received wages for hiring forty crossbowmen for the defence of Calais. Eighteen of them were paid 10*d.* a day and the rest were paid 8*d.* a day. In comparison, a standard foot archer at the time was paid only 6*d.* a day.[42]

The crossbow was a weapon beloved by Europe's secular elite, not one they feared. You would be far more likely to find a member of Europe's military elite using a crossbow to kill a peasant than the other way around. This myth probably has more to do with sixteenth-century authors, including most famously Cervantes in *Don Quixote*, complaining about firearms and the death of heavy shock cavalry than with the crossbow in the twelfth century.[43]

The Battle of Lincoln (1217)

The reign of King Henry III (r.1216–72) was a period of particular unrest in English history, some of it inherited from the troubled reign of his father King John and some of his own doing. During his 56-year reign, the longest of any medieval English monarch, there were two baronial revolts. The first was when he was just nine years old but was ultimately thwarted in less than a year. In contrast, for the second revolt he had a grown son of his own and briefly lost control of his kingdom until his son and future king Edward I (r.1272–1307) restored him to power. While that second revolt has several famous battles associated with it, it is the concluding battle of the first baronial revolt that is of particular interest to historians of the crossbow.

The battle in question was the result of a French invasion of England led by the crown prince Louis, later to be King Louis VIII (r.1223–6), with plans to usurp the throne from the very unpopular King John. Prince Louis landed his army in May 1216 and successfully marched on London. Louis and John engaged in minor skirmishes over the following months but John's sudden death in October 1216 completely changed the political calculus of the war. The young prince Henry was crowned king and William Marshal, Earl of Pembroke (1146/7–1219), was made the young king's regent and protector. Marshal was a famous warrior who at nearly 70 years of age was at the end of a very illustrious career that had begun during the reign of Henry II (r.1154–89), the new king's grandfather. The death of King John brought an end to many of the personal grievances felt by the nobility who now preferred the possibility of having a young king they could influence to a foreign ruler who might have his own ideas of how to rule England. This resulted in many nobles who had joined Louis switching sides and now supporting the young king against the French invaders.[44]

In May 1217, his alliances already at risk, Prince Louis divided his army. The force he led in person besieged Dover while the other, led by several prominent English rebels, moved to support another group of rebels from the north of England who were besieging the city of Lincoln. This second force successfully captured the city but failed to secure Lincoln Castle, which was built along the western walls of the city and had its own gate. The defence of the castle was led by Lady Nicola de la Haye (c.1150–1230) who was in her mid-sixties and had previous experience successfully defending castles. William Marshal assembled what forces he could and marched to relieve the castle, hoping that by delivering a swift victory he could crush the rebellion and push Louis out of England.

The scale of the two armies was relatively small with William leading less than a thousand men against a combined French and English force of around twice that number. Exact figures for William's army are debatable but the core seems to have been approximately 400 men-at-arms, many of them likely mounted, and just over 300 crossbowmen. In contrast, the Anglo-French army had probably over 600 knights and at least a thousand infantry. This made William's strategy something of a gamble, but on 17 May he committed and marched his army, wearing white crosses as a symbol of holy war, towards Lincoln. William initially approached Lincoln from the west, with his crossbowmen marching as the vanguard, to avoid a narrow river crossing where he could be bottlenecked. As he got closer, he repositioned to approach the city from the north-west which would have allowed him to either attack the northern portion of the city or else link up with the castle's garrison on the west.

The French commanders decided on a defensive strategy and instead of meeting William's army in the open field they took up positions along the city walls and continued the siege of the castle. This left William with a bit of a problem, as he had no siege equipment with which to carry out an extended attack on the city. Luckily for the royalist forces, some scouts discovered that one of the northern gates had only been partially blocked with rubble, and while the Anglo-French commanders thought it was secure it could easily be cleared for William's army. To support an attack on this gate, William's crossbowmen, led by Falkes de Breauté (died 1226), entered the castle through the western gate and took up positions on its walls to shoot down on the French inside the city. Simultaneously to Falkes' attack, William had Ranulf of Chester (1170–1232) lead a diversionary attack on the city's north gate, leaving William and his soldiers the exhausting job of clearing the north-western gate. This task took them several hours. Once the gate was open, the 70-year-old William led his forces charging into the city, much to the surprise of the Anglo-French defenders who were still trying to fight off the attack of Falkes and the crossbowmen in the castle.[45]

The French commander, the Count of Perche, fought a desperate and chaotic defence which ultimately led to his death, supposedly by having the point of a sword driven through the visor in his helmet. Soon after the count's death, Ranulf managed to break through the north gate and the French army was routed. After hearing of the defeat at Lincoln, Prince Louis broke off his siege at Dover and began negotiating for terms to end the conflict. When negotiations began to break down, a subsequent naval victory against French reinforcements sailing from Calais at the Battle of Sandwich brought the war to an end. Louis gave up any claim to the English throne in the Treaty of Lambeth in 1217, and William Marshal escorted him to Dover where the French prince sailed home.[46]

The Battle of Lincoln provides a fascinating glimpse into how crossbowmen were utilized in medieval warfare during this period. At the beginning of the conflict the crossbowmen marched a mile ahead of the main army laying down skirmishing shots and providing protection for the advance of the main body of the army. This seems to have been a common tactic for the deployment of crossbowmen at the time. Similar tactics were used by Philip II on his march to the Battle of Bouvines in 1214, as well as by Philip III (r.1270–85) during his march into Catalonia in 1285. This tactic was seemingly so common that when describing the battle, the contemporary chronicler Matthew Paris (c.1200–59) wrote that 'crossbowmen always lead the way'.[47] While it worked for William and others, it was not universally successful. The forces of the city of Liège deployed their crossbowmen this way against Duke John the Fearless and were defeated at Othée in 1408, but this did not lead to them abandoning the idea and as late as 1482 the crossbowmen of Liège mentioned in their ancient privileges that one of their primary functions was to go ahead of the main formations of the army and to cover its retreat in case of defeat.[48]

Once the battle was joined William's crossbowmen were able to break off as their own independent fighting unit, entering Lincoln Castle to attack the besieging forces. It can be easy to assume that archers and crossbowmen would only be deployed with the support of melee troops, but clearly in certain circumstances this was not necessary. While they had the castle to protect them from direct enemy assault, they were engaged in an extended battle with the Anglo-French forces before William's forces broke through the north-west gate.

The crossbowmen would also usually have carried some form of melee weapon with them to push back against attempts to take the castle by force. They were more than adequately able to defend their position against attack while simultaneously causing plenty of death and chaos.

The Battle of Lincoln shows the diversity of ways that medieval crossbowmen could be used in a battle, and while it should be noted that the relatively small scale of the conflict enhanced their overall impact, they were led by a seasoned commander in William Marshal and many of the tactics displayed at Lincoln were drawn from experience in larger conflicts and would still have been effective when used on a grander scale.

The Battle of Benevento (1266)

Few medieval battles have contemporary accounts that are as detailed as the Battle of Lincoln and fewer still make the same effort to describe the role of the infantry, archers and crossbowmen in the battle. The battles and sieges we have covered in this chapter are but a tiny fraction of the number of military engagements that took place during the High Middle Ages. The crossbow played an essential role in many more battles and sieges than we are aware of because in many cases the surviving sources neglect to describe their contribution. The Battle of Benevento is a great example of a battle where there are many surviving accounts, but historians still struggle to piece it all together.

The Battle of Benevento arose out of a particularly complex mixture of imperial, papal, French and Italian politics that all converged in one of the stranger sequences of events in medieval history. The battle has its origins in the disputes between the papacy and Emperor Frederick II (r.1220–50), who was also the king of Sicily. The papacy and the Holy Roman Emperors had a troubled relationship going back at least as far as the late eleventh century, and one of the papacy's great fears was that the empire would encircle them by uniting northern Italy, which officially belonged within the empire, and the Kingdom of Sicily, which included Naples and much of southern Italy. The papacy's fears were realised in the form of Frederick II and upon his death in 1250 the pope was not interested in seeing the problem return with his children.

Frederick II was succeeded by his son Conrad IV (1228–54), but Conrad died in 1254 and was succeeded in Germany by the anti-king William (1227–56) and by his son, also called Conrad but generally known as Conradin (1252–68), in Sicily before he was deposed by Conrad IV's illegitimate brother Manfred (1232–66). William would only outlive Conrad IV by two years, and what followed in Germany was a series of short reigns by foreign monarchs that would only fully end with the rise of the Habsburgs. That suited the pope, but in the meantime a descendent of Frederick II's Hohenstaufen family ruled Sicily. The pope shopped around Europe looking for someone to depose Manfred who would be susceptible to papal influence once they were on the throne of Sicily.

Many candidates were considered, including Edmund (1245–96), the younger son of Henry III of England, before the pope settled on Charles of Anjou (1226/7–85), the younger brother of Louis IX of France.[49]

While the initial agreement between Charles and Pope Urban IV (r.1261–4) was made in June 1263, it would be three years before Charles challenged Manfred in the field. In the meantime, Charles revised his agreement with Urban multiple times, with the increasingly desperate pope giving in to his demands as Manfred attempted to secure control over central Italy. Urban died on 2 October 1264 – casting the agreement with Charles into doubt. The election of a French cardinal as Pope Clement IV (r.1265–8) in February 1265 showed that the papacy had decided to put its faith in Charles against Manfred.[50]

In October 1265 Charles led a formidable army out of Lyons. Contemporary accounts claim his army numbered 6,000 heavy cavalry, 600 mounted crossbowmen and 20,000 infantry, half of whom were crossbowmen. These figures, especially the last one, were almost certainly exaggerations but it is clear that Charles had a large army and that a significant portion of it carried crossbows. The army was assembled with money from the papacy and drew soldiers from across France and Italy. It included many nobles as well as a substantial number of mercenaries, including both Pisan and Genoese crossbowmen.

The army made its way through the Alps and arrived outside Rome in mid-January 1266 where it was reunited with Charles who had travelled to Rome separately. Once united with his army Charles did not remain long in Rome, he advanced south with the hopes of bringing a swift end to the campaign. Manfred and Charles manoeuvred around each other, with Charles successfully taking several castles as he went, before eventually meeting on 25 February near Benevento in southern Italy.

Manfred's army included a mixture of Muslim cavalry and archers – one source claimed he had 10,000 Muslim archers and crossbowmen. These were drawn from a force of Muslim soldiers that Frederick II had settled in southern Italy for this very purpose: fighting against a papal-backed army. In addition to his Muslim troops Manfred had 1,200 mercenary German knights, a further several thousand Italian cavalry, and an even greater number of Italian foot soldiers, including some crossbowmen.[51]

The two sides mirrored each other in their deployment. Manfred lined up his Muslim archers as his first line, behind which he placed his elite German cavalry. The Italian troops and Muslim foot soldiers deployed on the flanks while Manfred's own bodyguard and a small force were kept in reserve under his personal command. Charles sorted his cavalry into three divisions and lined them up behind his foot soldiers and crossbowmen. The chronicler Saba Malispana (fl. thirteenth century) wrote that the battle began when Manfred's Muslim infantry attacked before the rest of their forces were fully prepared. After some initial skirmishing between the two sets of infantry the German cavalry charged and were met by a counter-charge from Charles' first cavalry group. When it looked like the Germans might triumph over Charles' Provencal cavalry, he committed his second division. Unfortunately for the Germans they had attacked too early and there was no relief force coming for them. Manfred's army had crossed a nearby river before the battle and this had delayed his deployment. Charles had his third cavalry group attack Manfred's army from the flank and the remaining troops broke and fled the field. Somewhere in the chaos of the battle Manfred was killed, securing Charles' total victory and new throne. His success would be relatively short lived, however, as in 1282 the Sicilian Vespers revolt would see him replaced on the Sicilian throne by Manfred's son-in-law.[52]

The frustrating thing about the Battle of Benevento is that we know there were thousands of crossbowmen present, but the accounts we have ignore them. Many of the chroniclers who wrote about the battle ignored the infantry entirely, describing the battle in terms of a clash between a few thousand cavalry. While the figures provided in chronicles are likely exaggerations and the armies were probably reduced from their starting sizes during the pre-battle campaign, there still had to have been thousands of crossbowmen and other archers on the field that day. This is the challenge for any historian attempting to reconstruct the use of the crossbow in the medieval warfare; there are far more battles like Benevento where little or no detail has survived than there are battles like Jaffa where we know almost every detail about how the crossbowmen contributed to the fight. The battles in this chapter have largely been selected because of the details that are available. So much could probably be learned about how the crossbow was used and how medieval commanders and soldiers viewed it if we had more detail about battles like Benevento.[53]

Chapter 6

The Crossbow in Late Medieval Warfare

The Siege of Orleans marked the point in the Hundred Years War when it became clear that the tide was turning against the English. The Earl of Salisbury began his opening bombardment of Orleans on 12 October 1428 and the siege would last until 8 May 1429 when Joan of Arc (c.1412–31) drove the English forces from the field and rescued the city on behalf of King Charles VII (r.1422–61). The story of the arrival of the Maid, how she convinced Charles of her potential, and how she rallied the forces of the French kingdom to defeat the English has been told many times in many different formats. In that story the Siege of Orleans is her crowning achievement, but it was also a moment of great personal danger.

The Siege of Orleans looked grim for the English almost from the start when on 24 October Salisbury was killed by a French cannon shot. He was replaced in command by John Talbot (c.1387–1453), a capable commander but who could not undo the greatest risk to the siege: the lack of manpower and funds. The king was not present at the siege: Henry V (r.1413–22) was dead and Henry VI (r.1422–61; 1470–1) still an infant, and Orleans was a massive city defended by over 2,000 French soldiers. The English had done their best to encircle it by capturing nearby fortresses and fortifying their camp, but it was to be a long siege and even before the arrival of Joan time was not on their side. Upon her arrival Joan immediately pursued an aggressive strategy – retaking several nearby castles and personally involving herself in attacks on the English defences. While she was ultimately successful, she also put herself at great risk and paid for it on a multiple occasions.[1]

Her first and most famous injury at Orleans was when an English archer shot her in the shoulder. There is some confusion around when and by what she was injured. The earliest account of her wounding in a letter by Lord Rotselaer dated 22 April 1429 where he recounts the arrival of Joan and that she predicted that she would be wounded by an arrow at Orleans. This miraculous prediction was repeated by Joan herself and was brought up again in her rehabilitation trial. Several other accounts of her wounding describe her as being wounded in an assault in May, presumably that of 7 May, which would be impossible if Rotselaer already knew about the wound when he wrote his letter. However, most accounts of the siege place Joan's arrival in early May, usually on the 4th, so either Rotselaer's letter has been misdated or it contains a genuine record of Joan's prediction about the future wound. The account of her fourth public examination during her trial in February 1431 describes the wound as having been from a crossbow bolt but given that Rotselaer's letter mentions an arrow and was likely written during the siege, that suggests an arrow was the more likely culprit. Joan did not let the wound get her down, the 1456 deposition of Louis de Coutes described her as only taking her armour off for long enough to be treated before donning it once again and rushing off to the front.[2]

The relief of the Siege of Orleans was Joan's greatest victory, vindication for the trust Charles had put in her, and the making of a legend. She suffered a similar wound during the Siege of Paris later that year, but rather than a great triumph this was one of the lowest points in her military career.

The Siege of Paris was very different from Orleans. Joan was no longer leading the flower of French chivalry to drive off an English army that was besieging one of the kingdom's most important city. Instead, it saw Joan on the offensive trying to retake the Anglo-Burgundian held city that had a troubled relationship with the French monarchy. Charles VII had signed a truce with the Anglo-Burgundians, but it explicitly excluded Paris and a few other communes – one of which, St Denis, Joan had recently recaptured with the royal army. Charles and Joan, along with the Dukes of Alençon and Bourbon, the counts of Vendôme and Laval, and Marshals Gilles de Rais and La Hire camped near Paris at the start of September. Joan and her closest allies were housed in the village of La Chapelle.

On 8 September Joan, the Duke of Alençon and Marshals Gilles de Rais and Jean de Brosse Boussac marched on Paris to begin their attack on Porte Saint-Honoré, one of the main gates to the city. Charles had assigned Joan the task of leading the assault. She immediately, and impetuously, launched an attack on the city. This was a disastrous failure that saw many French deaths and Joan suffering a severe wound. The chronicler Enguerrand of Monstrelet (c.1400–53) writing in the mid-fifteenth century recorded that in addition to many dead French '. . . the *Pucelle* was very gravely wounded, and she remained the whole of the day in the ditches behind a small mound, until vespers, when Guichard de Chiembronne and others came to find her'.[3] *The Journal of the Bourgeois of Paris*, written c.1431, described the Parisian defenders insulting Joan from the walls before shooting a bolt clear through her leg and into the foot of the man behind her. This story is almost certainly an exaggeration, but the core is consistent across the many accounts of her wounding – the assault had failed, and Joan was incapacitated. Unlike with her wound at Orleans where Joan could strap on her armour and return to the fray, at Paris she was incapacitated and pulled from the front lines.[4] In the wake of this disaster, Charles chose to withdraw his troops and abandon the siege despite having an overwhelming advantage in numbers over the Parisian defenders. A wounded Joan, facing her first-ever defeat, was unable to seriously object and the siege was given up.[5]

Joan had recovered from her wound by October and quickly returned to campaigning. In May 1430, her military career came to an end when she was trying to relieve the city of Compiègne which was under siege by a joint Anglo-Burgundian army. She was captured there by the Burgundians, possibly by an enemy archer, who turned her over to the English, where she was put on trial and then burned as a heretic. Charles VII would make several more attempts to retake the city of Paris, including in 1430, 1432 and 1434. He finally succeeded on 13 April 1436 when the city opened its gates to his commanders, their surrender resulting from the Burgundians abandoning the English cause and reconciling with Charles in the Treaty of Arras in 1435.[6]

Joan frequently put herself at risk and suffered multiple injuries in her short military career as a result. Her two most famous wounds show how bows and crossbows could inflict injuries without killing their targets or even severely impacting their ability to participate in military campaigns. While Joan's role in the French army was certainly unusual, no one could argue that she did not pursue it with great energy and determination – two elements that would have been greatly reduced had she been permanently disabled by an arrow or bolt. In the case of Orleans, she did not let the wound stop her from leading from the front. Her wound could have imperilled her ability to wield a sword, a significant concern for most soldiers, but since she insisted on only carrying her banner and no weapon it was no great problem. The wound at Paris was a little more severe in terms of its immediate impact on her participation – it probably hindered her ability to walk – although with Charles' decision to end the siege so soon it is impossible to say whether Joan would have found her way back to the front eventually.

Joan of Arc's experience shows the continued threat of crossbows even after the invention of plate armour and the arrival of gunpowder weaponry. While plate armour offered ever-increasing levels of protection for those who could afford it, it was not infallible. Joan wore plate armour custom-made for her, and while it is possible that the missile that hit her neck may have found a gap in her armour the one that penetrated her leg would almost certainly have had to go through at least some metal first. Joan's case also shows that these wounds were not necessarily fatal and that individuals could and often did survive a missile wound and continued their military careers for many years after. While gunpowder weaponry arrived in Europe around 1326 in Joan's story, and in many of the campaigns and battles discussed below, we can see how the crossbow continued to thrive in European warfare long after the first appearance of its eventual replacement. In fact, it was not until the sixteenth century that guns finally began to drive crossbows from Europe's battlefields, and even that was a slow and uneven process. The Later Middle Ages was a time of great innovation in military technology and one that saw gradual shifts in how armies were assembled and wars were fought, but throughout it all the crossbow endured.

The Battle of Courtrai (1302)

The Battle of Courtrai, also known as the Battle of the Golden Spurs, in 1302 has the strongest claim to be the battle that ushered in the Later Middle Ages. It has long been seen as a turning point in medieval warfare – the first time a force composed almost exclusively of non-aristocratic foot soldiers routed and destroyed the mounted nobility and heavy cavalry that had dominated warfare until then. While its position as a historical fulcrum is debatable – it should be seen as a product of specific circumstances rather than a tactical paradigm shift – there is no doubt that it was seen as a major event at the time, and it had a significant psychological impact on both the defeated French nobility and the victorious Flemish burghers. It also includes a great example of how crossbows were often deployed by French armies at this time – and the way they failed to deliver a French victory can provide historians with a useful lesson about the crossbow's strengths and weaknesses.[7]

The Battle of Courtrai arose from tensions between the powerful cities in the County of Flanders and the Kingdom of France. The French King, Philip IV 'the Fair' (r.1285–1314) had been trying to exert greater control over the County of Flanders from the 1280s but had faced strong opposition from Edward I of England. Edward was similarly trying to pull Flanders over to his side, since it was a major destination for England's main commercial export: wool. In 1297–8 Edward launched an expensive and largely unsuccessful invasion of Flanders which resulted in a truce between the two kings that left the Count of Flanders, who had been an ally of Edward, imprisoned and the county free for the French king to exploit. Philip IV invaded Flanders in 1300 and forced the Count of Flanders to submit to his rule. Unrest against French rule increased once Philip left, and in May 1302 many of the cities, starting with Bruges, rose in revolt against the French king. The rebelling citizens massacred any French people that could be found in the city, which included 300 soldiers led by James of St Pol that had been sent there to negotiate with the rebels. In response, Philip sent an army to invade the county. The rebelling cities rallied their own forces and met the French army outside the city of Courtrai, modern-day Kortrijk, on 26 June 1302.[8]

The Flemish army numbered between 7,000 and 10,000 in total, the vast majority of which were infantry armed with the Flemish *goedendag*, a type of polearm that was between 3ft and 5ft in length and approximately club shaped but with a large spike at its tip. It was a simple weapon common in town militias that effectively combined the benefits of both a spear and a club. Estimates of other soldiers are hard to put together, but J.F. Verbruggen estimated that there were probably less than 500 crossbowmen and a similar number of knights in the Flemish army. Opposing them was a force of approximately 3,000 French knights supported by 4,000–5,000 infantry, including an unknown number of crossbowmen and soldiers armed with javelins.[9]

The Flemish forces began by laying siege to Courtrai. Knowing that a French relief attempt was inevitable they established a strong defensive position nearby from which they could still maintain the siege. They placed their backs to a nearby river and dug a series of water-filled ditches in front of their position to interrupt the French attack and prevent the French knights from outflanking them. The French army spent three days inspecting the field before deciding to attack the Flemish rebels. The first stage of the battle was a duel between both sides' crossbowmen. Robert of Artois (1250–1302), the French commander, hoped that his crossbowmen could disrupt the Flemish position – either forcing them forward onto more favourable ground or else pushing them back so he could more easily bring up his own soldiers and mitigate the challenge of the ditches. The French crossbows quickly succeeded in overpowering their Flemish counterparts but struggled to take advantage of their initial victory. Once the Flemish crossbowmen fell back behind their own ranks, the French crossbowmen could not reliably reach the rest of the army without crossing several water-filled ditches which would have left them exposed and open to a counter-attack by the Flemish. After he saw that there was little more to gain from this exercise, Robert ordered the crossbowmen to pull back and prepared the cavalry to charge.[10]

Unfortunately for Robert and the many French knights under his command, the cavalry suffered one of the most surprising defeats in medieval military history. The crossbowmen had been unable to disrupt the main Flemish forces and the cavalry charging over uneven marshy ground failed to deliver much of an impact when they reached the Flemish lines. As a result, they were hacked down in their hundreds, with reports of over a thousand French knights killed including seventy-five senior members of the French nobility, Robert among them. The victorious Flemish militias stripped the dead knights and retrieved 500 pairs of golden spurs which were displayed in a nearby church.[11]

The victory at Courtrai, which was followed soon after by another victory at the Battle of Arques in 1303, was to be short lived and in 1304 Philip returned to Flanders, this time leading his army in person. The French army took up a defensive position near Mons-en-Pévèle and this time the Flemish went on the offensive. The battle opened with the Flemish crossbowmen, now supported with pavises – large shields that rested on the ground and provided protection for archers – shooting at the French in hopes of provoking an attack. The French held their ground and fighting was halted for five days of ultimately unsuccessful peace talks. When fighting resumed the crossbowmen on both sides once again opened with an exchange of missiles but neither proved very effective and soon the main forces clashed in a series of attacks and counter-attacks. Despite one Flemish counter-attack nearly reaching King Philip's tent and almost threatening his life, the Flemish army were ultimately driven from the field and their commander, William of Jülich, was killed. Casualties were extremely high with thousands dead on both sides, but since Philip held the field and the Flemish rebels were now leaderless, the French monarch claimed victory. He was able to follow his victory at Mons-en-Pévèle with a string of successful sieges that eventually resulted in the Treaty of Athis-sur-Orge in 1305 which recognized Philip as suzerain of Flanders and concluded the Franco-Flemish wars in Philip's favour.[12]

The battles of Courtrai and Mons-en-Pévèle show some of the limitations of the crossbow. There was nothing new about the deployment of the crossbowmen in the front of the armies, we saw the same tactic over 200 years earlier as Hastings and it was used to great effect at Lincoln in 1217. After initially triumphing in the archery duel at Courtrai, the French crossbowmen were caught between the need to be able to harass the Flemish foot soldiers and the risk of being counter-charged by those same soldiers. This challenge arose from the effective use of terrain by the Flemish: by forcing the crossbowmen to cross several ditches to be within range of the main Flemish lines they could effectively isolate them from the support that would keep them safe. While it may seem pointless for the Flemish to have sent a mere 500 crossbowmen out to duel with the much greater French force, perhaps the point was to try and draw those crossbowmen forward to isolate them – a trap they avoided. In both battles the impact the crossbowmen had on the conflict seems to have been minimal, but without a better idea of how many crossbowmen were in each army it is impossible to say if this was the result of them performing poorly or whether there were just not enough for them to change the outcome of either battle.[13]

The Battle of Crécy (1346)

It is impossible to write about the history of the crossbow without discussing Crécy. The battle, which took place in 1346 in north-eastern France, was one of the greatest triumphs of King Edward III of England (r.1327–77) and ensured the continuation of what is now called the Hundred Years War. Just as at Courtrai, many French nobles died at Crécy and the disorder that followed prevented the French from mounting substantial opposition to Edward III's subsequent siege of Calais. While they would recover for the next campaigning season, Calais would remain in English hands until 1558. The later Battle of Poitiers can claim the even more dramatic result of the capture of the King of France but given the broader context of the early stages of the Hundred Years War, Crécy was a more important turning point that ensured the war became the sprawling conflict it is known for today instead of yet another thwarted attempt by the King of England to reclaim lost territory in France.[14]

To an earlier generation of historians Crécy marked the arrival of a new form of English tactics, dominated by the longbow, which would define warfare for the next hundred years and bring the English many of their most famous victories. The battle opened with a duel between the English longbowmen, and mercenary Genoese crossbowmen hired by the French. The disastrous performance of the Genoese on 26 August 1346 has often been used as a way to critique the performance of the crossbow in general – especially when compared to the mighty 'English' longbow. It is worth reconsidering both the performance of the Genoese and their crossbows that day, as well as whether it is reasonable critique them based on the result of a single battle.[15]

The battle

Entire books have been written about Crécy and its many aspects. It was one of the most famous battles of its age and virtually every contemporary chronicler needed to provide their own take on it, some of them directly at odds with what another author was writing at the same time, which makes untangling the exact events and tactical dispositions of the battle a monumental task for modern historians. For example, most of the contemporary English accounts barely mention the role of the archers – and some do not mention them at all! In their book on the battle Andrew Ayton and Sir Philip Preston devoted over sixty pages to the many fourteenth-century accounts of the battle and the Gordian Knot they present to historians trying to piece together a coherent account of the battle. This book will not attempt to replicate that work, instead it will take a high-level approach to the account of the battle, but readers should bear in mind that the full story of the Battle of Crécy is one of staggering complexity and nuance.[16]

The Crécy campaign began in July 1346, when Edward III landed the largest army he would ever bring to France at the port of La Hougue in Normandy. Edward's landing was unopposed because the Duke of Normandy and many of the local nobility were besieging another English force elsewhere in France. Edward's army rampaged across Normandy engaging in multiple skirmishes, including capturing the city of Caen, before crossing both the Seine and Somme rivers and taking up a position near the town of Crécy-en-Ponthieu. King Philip VI (r.1328–50) brought his own army into Normandy and mirrored the English march, eventually pursuing them across both rivers.[17]

Philip's army arrived near the English camp at Crécy on the evening of 26 August, having marched all day while the English rested. In several accounts of the battle the French army did not wait until the morning and instead they began their attack on the English position in either the late afternoon or early evening. Jean le Bel (1290–1370), who spoke to participants of the battle, blamed the French attack on an undisciplined army and an unruly nobility refusing to back down from a fight once they could see their enemy before them. However, not every chronicler agreed with this version of events. Geoffrey le Baker (died c.1360), claimed that the French king was too cowardly to fight that evening and the battle took place on the following morning instead. Meanwhile, Giovanni Villani (c.1276–1348), whose account was written the soonest after the battle, described the French arriving at Crécy during the hour of Nones, which was approximately 15.00. Since they still had to deploy their forces for battle the actual fighting would have happened at least an hour later, which would put the battle sometime in the late afternoon.[18]

Regardless of what time of day it happened, the French seem to have engaged the English before they were ready and suffered a crushing defeat. King Philip was forced into flight – and according to some English accounts may have been wounded in the face by an arrow, although no French sources reported this story. At the end of the day the French dead numbered in the thousands, including the King of Bohemia and many other foreign allies who had come to support the French king.[19]

In the opening moments of the battle Philip ordered the Genoese crossbowmen to advance ahead of the main French position and begin shooting at the English, a similar tactic to that used at Courtrai 40 years earlier. This was probably an attempt to break up the English position, either in hopes of forcing them to attack or else loosening up their formation in preparation for a charge by the main body of the French cavalry. The Genoese crossbowmen and English archers exchanged missile fire but the Genoese, suffering significant casualties, fled the field. The cause of so many Genoese casualties differed between accounts of the battle, some blamed the English arrows, others reported that the French nobility became frustrated at the sight of the fleeing mercenaries and charged through them, massacring many as they went. Some versions even had them suffer both misfortunes.[20]

Why did they fail?

The need to explain the French defeat was a common theme across many continental chroniclers in the immediate aftermath of the battle. Most of the blame fell on the French leadership – both in the form of specific criticisms of the tactics and decisions those commanders made as well as a more general blaming of sinful lifestyles and arrogance which brought about ruin which accompanies nearly all accounts of military defeat in the Middle Ages. Explaining the failure of the Genoese mercenaries specifically was not the highest priority for contemporary writers, but several did try to explain why they fared so badly in their confrontation with the English archers. This thread of discussion expanded significantly as modern historians began to increasingly emphasize the role of the English archer in the victories of the Hundred Years War – which was accompanied by a tendency to compare the 'yeoman archer' with the 'elite mercenary' Genoese. The combat between the Genoese and the English at Crécy has become a proxy for debates about the strengths of the longbow in comparison to the crossbow – a debate that is tied up in English national identity and the claim to the longbow as an essentially English weapon. Further aspects of this debate will be discussed later in this book, and for the moment this section will focus just on Crécy.[21]

Two accounts, the *Grandes Chroniques* and Jean de Venette's *Chronicle*, explained the defeat of the Genoese by recording that there had been a shower of rain right before the battle and that this rain had ruined the crossbows' strings, thus rendering them ineffective. Supposedly this affected the crossbowmen because their weapons could not be easily unstrung while the English archers could unstring their bows and hide the strings under their caps. This story is an unlikely explanation for a number of reasons.[22]

Firstly, Jean de Venette claimed that the rain caused the crossbow strings to tighten, but modern historians have pointed out that it is more likely that the rain would have caused them to loosen and stretch instead. Crossbow historians have also emphasized that strings were likely to have been treated with a layer of fat, which would render them resistant to moisture, and that crossbows – especially composite ones whose glue would be susceptible to rain damage – would likely have been kept covered during rainfall. Possibly had rain fallen during the battle that could have caused problems, but it is very unlikely that rain before the battle would be allowed to damage the crossbows. Even if none of these precautions were taken, a slight loosening – or even tightening – of the crossbow string would not be enough to render the weapon unusable and would not be sufficient to explain the disastrous result for the Genoese.[23]

Another slightly more plausible story of rain was reported by Antonimo Romano, who claimed that a shower of rain had turned the ground around Crécy into a muddy mess which rendered the crossbowmen unable to quickly span their weapons. He wrote: '. . . when the Genoese tried to load their weapons by putting a foot in the stirrup, the foot slipped, and they could not plant it on the ground.'[24] While it still has the same challenge that the rain is only mentioned in a handful of accounts, this is an example of how it very well could have impaired the performance of the Genoese. It is also worth noting that while Giovanni Villani did include a brief shower of rain that delayed the fight, he did not mention anything about it affecting the crossbowmen's performance. Given that Villani wrote his account very soon after the battle and may have interviewed members of the Genoese crossbowmen, the absence of this detail from his account is telling.[25]

A more generally-agreed explanation for why the Genoese mercenaries failed at Crécy was that when they marched forward to fight the English, they did so without their full equipment. If we accept the more likely timeline that the battle took place soon after the French arrived at Crécy, then the baggage train was probably still arriving on the field when the Genoese deployed. Included in that baggage train would have been at minimum the crossbowmen's pavises, as well as potentially most of their ammunition and any armour they had not worn on the march. If this were the case, then the Genoese would very easily have been outmatched in a duel of archery as they would have lacked any protection at all from the onslaught of arrows. It could also explain their apparent lack of enthusiasm during the battle which caused their early retreat. After marching all day and then being forced to advance into enemy shooting without the proper equipment one could hardly blame the mercenaries if they decided that they were not being paid enough to put up their lives in this kind of danger, so maybe they put on a show of sending off a few shots before trying to get out of there alive.[26]

The matter of rate of fire is another popular explanation, one that emphasizes the strengths of the longbow rather than the disadvantages of the Genoese. In this version, the superior rate of fire possible with a longbow was just too much for the slower-shooting crossbowmen to endure and they were forced to retreat. In an interesting and highly speculative article, Russell Mitchell attempted a case study on the possible importance of rate of fire in medieval warfare in general by using the example of Crécy. He argued that the English could only have maintained a high rate of fire at the cost of only being vaguely accurate, but if they were firing into a packed formation of Genoese soldiers, then accuracy would have been of minimal importance. Mitchell argued that had the Genoese brought out their pavises as well as their helmets and other armour, the massed volley of arrow fire would have been ineffective, since the inaccurate shots would not have been lethal or disabling. In response, the English would have had to take better aim with each of their shots in order to be effective, which would have greatly reduced their rate of fire. While highly speculative, Mitchell's article made good points about how this debate often takes the longbow's rate of fire as innate without examining the costs that would have come with actually sustaining it. Further challenges to maintaining a high rate of fire with the longbow would have been the physical endurance of the archers – shooting a bow all day would have been exhausting – and the limitations on the amount of ammunition they brought. Bert Hall noted that if we accept that the English shot upwards of half a million arrows at Crécy, that would still only have been roughly ninety arrows per archer for a battle that lasted upwards of six hours, which would have been only fourteen arrows an hour if they shot at a steady rate. The longbow could certainly unleash a burst of arrows at a critical moment, and they may have done that to the Genoese at Crécy, but it must be remembered that this would have come at the cost of sustained fire later in the battle.[27]

Lastly, there is another far simpler explanation for why the Genoese performed so badly that day. Crécy was a battle marked by French disorganization and confusion. Historians should not assume that the chaos in French leadership and tactics was limited to only the elite cavalry. Chaos in the leadership would have extended to the whole army. As a professional mercenary company, one could expect a greater level of discipline from the Genoese, but even well-disciplined troops will suffer when orders from command are confused and when the rest of the army is in disarray. This explanation is also something of a warning: historians should be careful before using the events at Crécy to explain late medieval tactics in general. At least for the French it seems to have been a particularly bad day and it would not be entirely fair to judge the performance of them, the Genoese or the crossbow based on one incident. Supporting evidence for this can be seen in the fact that the Genoese crossbowmen continued to be an important element of European warfare. The fact that they had been so soundly defeated at Crécy was not seen as a reason to stop hiring them. In fact, Edward III himself hired forty Genoese crossbowmen as part of the Calais garrison in the 1370s. These crossbowmen were defending the very city that Edward had taken as a result of his victory at Crécy. It can be tempting when looking back to assume that Crécy shows the obvious superiority of the longbow over the crossbow that contemporaries must have missed. However, remember that medieval commanders knew far more about the business of waging medieval warfare than any modern commentator looking back ever will and if they still saw value in the services of the Genoese and their crossbows then it must have been there.[28]

The Battle of Aljubarrota (1385)

King Ferdinand I of Portugal (r.1367–83) died in 1383 with his only his daughter Beatrice as heir. Beatrice was married to King John I of Castile (r.1379–90), and her succession would have resulted in a de facto annexation of Portugal. In opposition to King John, the nobles of Portugal selected an illegitimate son of the previous king, Peter I (r.1357–67), and elevated him to be King John I of Portugal (r.1385–1433). Unwilling to give up his claim, John of Castile invaded Portugal in 1385. He had the support of the French King Charles VI (r.1380–1422), and his army probably numbered 20,000–25,000 soldiers including 2,000 mounted French men-at-arms. John of Portugal had support from England and received 640 English mercenaries, an even mix of men-at-arms and archers, to reinforce his army which probably totalled around 10,000 men.[29]

John of Castile marched his army towards Lisbon, knowing that if he managed to take the city, he could bring a quick end to this rebellion. John of Portugal moved to block his advance by taking a position north of the city near Porto de Mos. Thanks to recent archaeological excavations we know the site of the battle, including some of its geographical features. John of Portugal first took up a position at the north end of a small hill on the road between Leiria and Alcobaça but was forced to reposition to a few kilometres to the south when John of Castile tried to circumvent him. John of Portugal may have planned for this event, however, as he quickly fortified this new position with a ditch and barricades made of trees chopped down from a nearby forest. Unable to escape from his rival, John of Castile was forced to give battle on 14 August 1385.[30]

John of Portugal deployed his army in a recognizably English fashion. He positioned a core of his dismounted men-at-arms in a central battle while on the wings he placed his archers – the 300 English archers mixed in with at least 800 crossbowmen. The archers were protected by constructed defences and a small detachment of men-at-arms. The Portuguese king had arranged the battlefield such that the Castilian army would have to advance through a narrow gap to reach him while the archers rained shots down on them. This tactic worked exactly as planned and the much larger Franco-Castilian army charged into a death trap with disastrous results. King John of Castile escaped the defeat because he was too sick to participate in the front lines and was carried around on a litter, although his illness and the contentious relationship between the French and Castilian men-at-arms may have greatly contributed to his army's defeat.[31]

At Aljubarrota John of Portugal showed that a combined force of crossbows and longbows could be used to effectively replicate the successful defensive tactics of the English in the Hundred Years War. There are a few interesting differences between Aljubarrota and more famous battles like Crécy or Agincourt that show the versatility of these tactics. For one, John of Portugal's archers made up no more than a tenth of his overall soldiers. This is a sharp contrast with the army of Henry V at Agincourt which had a ratio of at least three archers to every man-at-arms. This shows that these victories could be replicated with smaller ratios of missile weapons. Another interesting difference is that while many famous English battles raged for most of a day, Aljubarrota was determined in the opening hours – one account claims within the first half-hour but this must be an exaggeration. Aljubarrota showed how crossbows could be used as a vital part of this strategy, and it was not alone. In 1387 the famous English mercenary captain John Hawkwood (c.1323–94) achieved a similar victory with a combination of longbows and crossbows on the wings at the Battle of Castagnaro.[32]

The Crossbow at Sea

During the thirteenth and early fourteenth centuries archery, and the crossbow in particular, were omnipresent in naval warfare. The range of ships that were used in naval combat narrowed during this period and naval warfare became more standardized. The most common ship of the era was the galley, a primarily rowed vessel that usually also had a sail for occasional use. The most common form of galley was a bireme, meaning one that had two benches of rowers, but other forms with more or fewer benches were also used. A standard bireme had approximately 108 rowers supported by a complement of 20 to 30 crossbowmen and marines. Galleys were fast and manoeuvrable, but they required large crews to operate effectively.[33] Laws dictating the equipping of galleys put in place by Venice and other Italian city states show the popularity of the crossbow in these ships. Initially crossbowmen were expected to act as oarsmen as well, but over time it seems that they were spared this duty and allowed to concentrate on their weapons. Venetian records indicate that the composite crossbow was preferred and that it came in two types: the crossbow with a stirrup and the crossbow *de torno vel de pesarola*, probably meaning a windlass crossbow.[34] The cog was a single-masted ship with a square-rigged sail and a large, rounded hull. It required fewer crew than the galley which in turn left more room for soldiers but at the cost of mobility. In the fourteenth century the cog was supplemented by the carrack, a larger three-masted ship with a similar rounded hull. The military role of cogs and carracks could be further enhanced with the addition of wooden towers, called castles. Castles elevated archers and siege weapons and allowed them to rain death down on enemy ships. Both the cog and the carrack were difficult to board from a galley since the latter ship sat low in the water and the former both had high sides – a factor that was important since one of the primary goals in medieval naval warfare was to board and capture enemy ships. However, the galley could easily outrun and outmanoeuvre both types of ship.[35]

One of the most active theatres for naval warfare was the English Channel during the Hundred Years War. Among the most interesting of the many battles fought in or near the Channel was the Battle of Winchelsea. In the summer of 1350, a fleet of ships from the Kingdom of Castile, an ally of the French king, sailed into the port cities of Flanders. Edward decided to punish the Castilians for sailing through his waters and assembled a fleet, including at least twenty-five royal ships, at Winchelsea on England's southern coast. Since Edward also controlled Calais, he could monitor the fleet from both sides of the Channel and attack when it tried to sail home. While the Castilian fleet was comparable in terms of its total number of ships, it included significantly more carracks than the English had, which made it a daunting foe.

On 29 August 1350 Edward personally commanded his fleet as it intercepted the Castilians. The contemporary chronicler Geoffrey le Baker described the battle thus:

> Finally about the hour of vespers on the feast of the beheading of St John the Baptist the two fleets clashed. The huge Spanish galleons towered above our own galleys and ships like castles above cottages. They made frightening attack on our men, repeatedly inflicting deep wounds on many of them by rocks which flew down from the mast turrets and by quivering bolts and arrows. None the less our men kept up the fight at close quarters with spears and swords and bravely defended themselves with the weapons the fleet possessed.[36]

In Geoffrey's account the Castilian crossbowmen were forced back by volleys of arrows from the English longbowmen and in the ensuing melee the English successfully captured about half of the Spanish fleet. While Geoffrey painted this as an unmitigated triumph, and the captured vessels no doubt contained substantial wealth, it came at a great cost. The English suffered horrific casualties – many ships were reduced to skeleton crews by the time the fighting ceased. The half of the Castilian fleet that escaped ultimately found itself back in Flanders due to unfavourable winds but this time they were unopposed by the English – another sign that while Edward might theoretically control the Channel, the casualties he had suffered limited his ability to enforce it.[37]

The triumph of the English longbowmen over the Castilian crossbows, especially given the latter's favourable position and larger ships, would suggest that the longbow was the superior weapon for naval combat. However, while at Winchelsea the English effectively used their higher rate of fire to suppress the Castilian crossbowmen, that would not always have been an option. Naval battles generally did not allow for an easy retreat and could rage on for hours on end – the Battle of Sluys in 1340 (Plate 28) continued even after nightfall. Winchelsea by contrast was relatively short – according to Geoffrey it only began in the evening and the English were trying to intercept a fleet on the move, which would have allowed the archers to safely shoot all of their arrows at full speed without worrying about running out. It also must be stressed that much of medieval naval warfare was still fought up close and personal via boarding actions. Archers and crossbowmen provided essential support and could inflict serious casualties on the enemy, but they alone could not decide a fight.[38]

The Crossbow in the Northern Crusades

The Northern Crusades, a series of campaigns to conquer and colonize northern and eastern Europe linked primarily by religious incentives given to participants by the Church in Rome, had their origins in the twelfth century but they really escalated in the thirteenth and fourteenth centuries before slowly coming to an ignominious end in the fifteenth and sixteenth centuries. Warfare in the Baltic and its neighbouring regions was very different from the rest of Europe, driven in a large part by the environment. This region was still heavily forested and difficult to march through, meaning that early campaigns were almost entirely done by ship, with invading Crusaders sailing up rivers to raid and settle. The campaigning seasons for overland invasions were relatively short, with a brief period in spring and a theoretically longer period in winter usually limited by available supplies to feed an army. While the crossbow was used during several Christian invasions of the Baltic it was with the arrival of the members of The Order of Brothers of the German House of Saint Mary in Jerusalem, better known as the Teutonic Knights, that it became a staple of Baltic and Eastern European warfare.[39]

The crossbow was written into the practices of the Teutonic Order at a fundamental level. One of the lines in the statutes of the Teutonic Order read: 'The marshal may receive from the crossbow workshop for the brothers' smaller crossbows for drawing with the feet or bows, according as he sees crossbows or bows to be necessary for them.'[40] The order took an archery-heavy approach to warfare, similar to English armies in the Later Middle Ages. Initially the Teutonic Order often recruited armies where half to one-third of levied troops were equipped with crossbows. After Lithuania officially converted to Christianity in 1387, the order became increasingly dependent upon mercenaries due to a decrease in new recruits and lower Crusade participation generally, but this did not decrease the number of crossbowmen in their armies. There is clear evidence of elite crossbow soldiers being a common pool from which the Teutonic Order recruited to make up for their shortfall. The grand master of the order sent a letter to Philip the Bold, Duke of Burgundy (1342–1404) in which he offered his thanks for a gift of wine and the crossbowmen that the duke had sent to fight with the order. The grand master reported that the crossbowmen had been well behaved. In 1409, Grand Master Ulrich of Jungingen (1360–1410) hired Genoese crossbowmen for his impending war with Poland-Lithuania, but the record does not make it clear if these mercenaries arrived in time to participate in the disastrous Battle of Tannenberg where the order was utterly crushed by the combined Polish-Lithuanian army – a defeat the knights never fully recovered from and which made them even more dependent on mercenaries to fight their wars.[41]

The Teutonic Order maintained a large supply of reserve crossbows and crossbow makers tasked with maintaining and replacing the weapons as needed. In 1396, the commander of Ragnit deposited 1,150 pieces of wood to be made into crossbows with a citizen in Danzig on behalf of the grand master.[42] This was in addition to an even larger deposit of wood to be turned into normal bows, showing that while fond of the crossbow the order used the bow as well. They were not entirely dependent upon outside help to supply their crossbows. The order maintained numerous workshops, called *Schnitzhäuser*, in their great castles. The *Schnitzhäuser* in Marienburg reportedly had 1,200 billy goat horns and 26,000 sinews for making composite crossbows in 1409. They also had enough birch bark to cover 1,200 crossbows. Bark coverings were an important method of protecting composite crossbows from the elements. At the beginning the fifteenth century the order had at least 4,167 crossbows in their castles in Prussia – and presumably even more assigned to brothers for active use. Of these crossbows at least 1,496 were stirrup crossbows and at least 66 were spanned by a *cranequin*. In addition, they had over a million bolts for using with those same crossbows – a figure that seems ridiculously large until you realise how many bolts could be used in an active siege. In 1431 the commander of Rehden sent a message complaining that he did not have enough bolts to withstand a serious attack – the commander had only 1,800 bolts. That this figure was felt to be too low gives an idea of how many a commander would require to feel safe.[43]

How did the Teutonic Knights and their allies put these crossbows to use? The form of warfare in the Baltic and Eastern Europe changed over the centuries. In the earliest phases, approximately from the twelfth to the early fourteenth century, the purpose of the Crusade was to carve out a section of

Christian-controlled territory from which further Crusades could be launched. The easiest point of entry into the region was down its many rivers. However, natives of the Baltic region used those rivers, often constructing fish weirs across them which restricted access to ships and provided necessary food for the local population. The Crusaders needed to break through these weirs if they were to penetrate deep into Baltic territory and the local populations were not interested in letting them. It was in these battles over the rivers that the crossbow played its first significant role in the conflict. Crossbowmen could lay down barrages of missile fire against local warriors while the men-at-arms and other soldiers destroyed the weirs. It was better for the Crusaders to keep the enemy at bay while they destroyed the weirs than it was to fight every single commune in a pitched battle on their way up the river – even if they were likely to win those battles, they would be very time consuming, and the campaigning season was short.[44]

Once the Crusaders had succeeded in seizing territory the next challenge was defending it. This was another situation in which the crossbow was a valuable tool in their arsenal. While stone and brick castles would become more common as the Teutonic Order became more established, the early days of the Crusade were marked by the hasty construction of simple and sometimes flimsy wooden forts. Having a strong garrison of crossbowmen, along with the occasional catapult or *ballista*, gave the Crusaders a significant advantage in holding their forts against counter-attack until the next campaigning season brought more Crusaders to expand their territory further.[45]

Slowly the Crusaders, and especially the Teutonic Order, carved out territories for themselves and became entrenched in Eastern Europe. This process was not universally successful, the invaders being defeated on several occasions. They established semi-permanent footholds in Prussia and Livonia but struggled to establish a long-term hold elsewhere. Their permanent settlements also fundamentally changed the nature of their Crusades. As they conquered more territory, they had to travel further each campaigning season to reach new lands to invade – something that was very challenging given the short campaigning season. For the Teutonic Order this meant relying even more on armies composed primarily of cavalry since only they could travel fast enough to reliably outpace the turn in the weather that would mark the end of the season. This also resulted in a decrease in the role of crossbows in the offensive, although as we have seen above that did not mean the Teutonic Order did not continue to use vast numbers of crossbowmen to hold their many fortresses and cities. This is not to say that the crossbow entirely disappeared from the battlefield. For example, Gdansk sent sixty soldiers including thirty crossbowmen on Crusade in 1405, but their role was diminished from what it had been during earlier phases. By the late fifteenth and early sixteenth century the crossbow was slowly being phased out in favour of gunpowder weapons.[46]

While the crossbow was predominately used by the Teutonic Knights and other Crusaders, it was not limited to them. The first Grand Duke of Lithuania, Mindaugas (r.1236–53), had Lett auxiliaries that were equipped with crossbows they had acquired from the Sword Brothers, a group of Crusaders who would later be reformed as the Livonian Order. Recent excavations of Vilnius Castle have uncovered substantial evidence that the crossbow was used in its defence during the late fourteenth and early fifteenth centuries. Crossbow nuts, horn inlays and the remains of bolts within the castle all clearly point to its use by the Lithuanians in the castle. This means that very likely crossbows were used by both sides during the five attempts by the Teutonic Order and its allies to take the castle between 1375 and 1394. There is further evidence that suggests there may have been a crossbow workshop in the area of the castle, meaning that the Lithuanians were making and repairing their own weapons instead of importing them from abroad. Another Grand Duke, Prince Jagiełło (r.1377–81; 1382–1434). used mercenary crossbowmen when he defeated the Teutonic Knights at Tannenberg in 1410. It seems likely that the Teutonic Knights had few or no crossbowmen with them as their army was almost exclusively cavalry, while Jagiełło had a force of mounted crossbowmen that he had dismount to provide missile support.[47]

The crossbow's role in the Northern Crusades shows the great geographical reach of the weapon as well as its versatility. While many of the principles of warfare in the Baltic and Eastern Europe were similar, defending forts or providing supporting missile fire for a primarily melee army, the region also included many unique challenges that soldiers of both sides had to consider when launching a campaign or defending against an invasion. In these differences we can see how the crossbow could be used in different ways as well as its continued popularity with the soldiers and commanders of medieval Europe.

It is also worth bearing in mind that while this section has taken a largely linear narrative, in practice the history of warfare in the region and the deployment of the crossbow followed an overall much messier timeline. The phases of the conflict roughly align with what has been outlined above, but within each phase individual campaigns could come in many shapes and forms – some of which may seem more appropriate in a later or earlier phase of the conflict. History is never simple nor particularly linear.

The Hussite Wars

'Then Žižka marched out of Kutná Hora to fight the King, and he built his encampment with his war wagons in the field before Kutná Hora and there they battled all day long.'[48] That is how the author of *The Very Pretty Chronicle* described the famous tactic of the Czech Hussite general Jan Žižka (d.1424): the *wagenburg*. Before we can address the *wagenburg* it is necessary to provide some context to who the Hussites were.

On 6 July 1415 the religious reformer, preacher and former rector of the University of Prague Jan Hus (c.1372–1415) was burned as a heretic on the orders of the Council of Constance. This was despite Hus having been guaranteed a safe return to his home in Bohemia by King Sigismund of Hungary (r.1387–1437). Understandably, the Czech followers of the reform movement that Hus was at the forefront of were not pleased by this turn of events. Matters only deteriorated further when Wenceslas IV, King of Bohemia (r.1378–1419), died childless and was succeeded by his brother Sigismund, the same man who had allowed Hus to be killed.

The religious reform-minded people of Bohemia soon rose in revolt against their new ruler, demanding acknowledgement of the right to practice their religion according to the reforms put forward by Hus and his fellow theologians. These principles were outlined in the Four Articles of Prague which demanded that the right of the Hussites to preach their doctrine in Bohemia be recognized. They also wanted the right to give Communion in both kinds – i.e., bread and wine – to all members of the congregation, the removal of earthly wealth and possessions from the religious classes, and both the stricter punishment of mortal sins and the restoration of the good name of Bohemia from the slanders that it had been levelled against it. Sigismund was slow to react to his acquisition of the throne of Bohemia and the unrest that had come with it, and when he did arrive was entirely unprepared for the effectiveness of resistance that the Hussites organized against him. Central to that effective resistance was the brilliant tactician Jan Žižka.[49]

Žižka's origins are surprisingly difficult to pin down. We know he was from Trocnov in southern Bohemia. He was a member of the lesser nobility and held the rank of squire. He served for some time as a mercenary, including in the service of the King of Poland. It is even possible that he was part of the Polish force that defeated the Teutonic Knights at Tannenberg, although it is impossible to say if he fought in the battle or was only providing logistical support. His exact age is unknown, but by the time of the Hussite rising he would have been at least 50. What we do know about the one-eyed general is that he was a highly effective commander who delivered a string of victories for the Hussite movement that eventually allowed them to secure the religious privileges they desired, even if his personal faction, the Taborites, was largely wiped out after his death. One of the key's to Žižka's success was his tactic of the *wagenburg*.[50]

A *wagenburg* was a group of armoured war wagons laid out in a rough circle with their wheels chained together to prevent them from moving. The gaps between wagons were generally used either as artillery emplacements with medium to large cannons placed there, or else they were covered by a pavise, sometimes with a smaller gun mounted behind it on an elevated platform. Each section of the *wagenburg* was usually defended by between fifteen and twenty soldiers armed with a variety of weapons. An average *wagenburg* party would be equipped with at least six crossbows, two handguns and an assortment of basic melee weapons including spears, halberds, clubs and repurposed farming flails. Logistically, wagons were often grouped into rows with each row consisting of sixty wagons. In 1423 Žižka campaigned with a force of 'only two rows of wagons', or 120 wagons in total, which should give some indication of the scale of the *wagenburg* armies when they deployed in force.[51]

Hussite *wagenburg, Cod. 3062, Fol. 148r: Sammelhandschrift Zur Kriegskunst,* Österreichische Nationalbibliothek, Vienna. (Public Domain image, Wikimedia Commons)

The *wagenburg*'s ability to facilitate the deployment of gunpowder weapons against armoured cavalry has generally been the aspect that has drawn the most attention. Guns were very slow to reload and thus very vulnerable to cavalry charges, a fact which largely limited them to siege warfare during the Middle Ages. However, Žižka showed how guns could be used to devastating effect on the battlefield. That said, it is an oversimplification to associate the *wagenburg* exclusively with guns. Guns were just one weapon in Žižka's arsenal. When describing the Hussite *wagenburg* a member of the Teutonic Order wrote that crossbows and the best gunpowder weapons available were equally effective, meaning there was no reason to prefer gunpowder. Žižka seems to have agreed since the average *wagenburg* had three to four times more crossbows than it did portable firearms.[52]

A standard deployment of Žižka's army included multiple *wagenburgs* on favourable terrain, usually on a hill. The tactic generally relied on Žižka's ability to force his enemy to attack him since his army was largely immobile during battle, but at the same time the wagons allowed him to move his fortifications to the ideal position to force his enemies to attack him on unfavourable terms. At the Battle of Porici in 1420, and probably at other times as well, Žižka further enhanced the defensive nature of his *wagenburgs* by ordering the digging of a ditch.[53] In some cases the *wagenburg* fulfilled the function of a fort. In 1424 near the town of Kostelec, Žižka's *wagenburg* was under siege by the forces of Prague for several days before he was able to flee in the night with his army and wagons intact.[54]

The *wagenburg* was also capable of surprising mobility when it was required. The battle outside Kutná Hora in 1421 showed how Žižka's army could switch from largely immobile to suddenly very fast in the course of a single day. At Kutná Hora Žižka deployed his army in the usual *wagenburg* formation in front of the city walls against King Sigismund's much larger army. While Žižka's formation was more than capable of resisting the attacks of Sigismund's army, a force managed to circle around past Žižka and in conjunction with treacherous elements in the city breached the walls and took control of Kutná Hora without defeating Žižka in the field. Finding himself encircled and without anything to defend, Žižka responded by leading his army to break through Sigismund's forces near where the king had positioned himself and escaped deeper into Bohemia. That Žižka took his *wagenburg* with him is made clear by references to him setting up his war wagons to defend against any pursuit by the German king on the following day. Žižka later managed to reform his army and drove Sigismund from Bohemia, retaking Kutná Hora in the process, and delivering the king's heaviest defeat of his many unsuccessful campaigns in the region.[55]

The *wagenburg*, and the crossbow's role in it, was in many ways an expansion of the defensive tactics seen at Jaffa in 1192. An even more direct parallel can be seen in the account of Crécy written by Giovanni Villani in which he claimed that the English deployed their supply wagons in front of their archers to protect them from the French. The difference was that Žižka added a greater level of sophistication to these tactics and used them repeatedly rather than during just one battle. The *wagenburg* was a new way to fight an old-fashioned style of war, a way for a clever general to ensure that he could fight a defensive battle to his own advantage whenever he wanted.[56]

Žižka's first recorded use of the *wagenburg* tactic was December 1419 when he drove off an attack from royalist forces near the city of Pilsen.[57] He successfully defeated several much larger armies using the *wagenburg*, including at Sudomer, Porici and Bor Panský in 1420, Vladař and Kutná Hora in 1421, Německý Brod in 1422, Horice in 1423, and at Malesov in June 1424. The tactic was far from static during this time. Žižka expanded the number of wagons in his army over the course of his career as well as incorporating cavalry and other soldiers to support the wagons. It should also be stressed that while the *wagenburg* was widely used throughout the Hussite Wars, the conflict also featured countless sieges and other forms of more traditional medieval warfare, at which Žižka and his compatriots proved to be more than capable.[58]

The *wagenburg* tactic outlived both Žižka and the Hussite wars. It would go on to be used in many conflicts across the European continent. It was famously used at the Battle of Cerignola in 1503. The Spanish general Gonzalo Fernández de Córdoba (1453–1515) deployed a *wagenburg* supported entirely by gunpowder weapons which drove off numerous charges by the French. Gonzalo probably would have achieved similar results had he used crossbowmen alongside his arquebuses, but that was increasingly an outdated technology. Cerignola showed the versatility and longevity of Žižka's tactical innovation, but it also showed that the writing was on the wall for the crossbow.[59]

The End of the Crossbow

The archer in Renaissance France

France fought most of the Hundred Years War without showing much interest in mirroring England's use of massed archery. The reasons for this have as much to do with how each kingdom recruited their armies as it did with their opinions on the role of archery in warfare. However, during the reign of Charles VII – the French king who ultimately drove the English out of all of France except for the Pale of Calais – the French monarchy began to take a very different stance on military recruitment and the use of massed archery. For the French, however, massed archery included both longbows and crossbows. The exact ratio of longbows to crossbows is impossible to determine since most sources were at best vague about the types of weapons used by the archers and many records do not mention it at all. Over the course of the fifteenth and early sixteenth centuries even the meaning of the term 'archer' itself began to change. The late medieval and early Renaissance

French armies show how the longbow and crossbow were not in opposition, but rather inextricably intertwined with each other. To explore the adoption of one is to consider the use of the other.

Charles VII was not the first French king to explore increasing the role of archery in his kingdom's armies. In 1367 Charles V (r.1364–80) had ordered that a register be made of every archer and crossbowmen in his '*bonnes villes*', meaning the cities of his realm, and in 1369 he banned the participation of the people in popular sports, including football, and instead forced them to practice their archery.[60] This ban was repeated in 1384 by his son Charles VI.[61] A very similar order had been made during the reign of Edward III in England and continued to be reissued by his successors. The 1371 accounts of the master of Paris includes payments for the erection of butts for crossbowmen to use as targets on the Ile de Notre Dame, as well as the repair of two butts for archers on the same island. In 1379 crossbowmen were mentioned as shooting at the '*champs des arbaletriers*', which was a field that ran along the walls of Paris.[62]

Charles V's efforts to promote archery seem to have been primarily concerned with the defence of his realm. This marks a difference between his project and that of his grandson some 60 years later, but the roots of that later programme can be seen in the efforts that both Charles V and Charles VI took in the wake of the disastrous Treaty of Bretigny in 1360.

The core of Charles VII's new military was to be the *Compagnies d'Ordonnance*, a standing army of sorts consisting of 12,000 men grouped into 20 companies. Each company was made up of 100 *lances fournies*. A lance was a type of tactical unit common in the Later Middle Ages, although with its composition varying by region. Charles' lances consisted of a man-at-arms, a *coustillier* – a soldier armed with a large knife called a *coustille* – with his page, two archers and either a *valet de guerre* or a third archer. All members of a lance were to be mounted although they would not all necessarily fight on horseback. While the ordnances were open to both types of archer, in practice a majority of them carried crossbows. An important factor in the eventual success of the lances was that Charles took great care to ensure that they were both well-equipped and adequately paid, two of the most important elements for the success of any army.[63]

The usual ratio of archers to men-at-arms in French armies had been two men-at-arms for each archer, but a lance had the reverse ratio which meant Charles had to find a massive supply of archers for his armies. To this end Charles made efforts, briefly mentioned above, to promote archery in France. In 1437 he confirmed the privileges of both the fraternity of the crossbow and the fraternity of the longbow in Paris. In 1445 he granted permission for the establishment of a fraternity of archers in Tournai and in 1446 he confirmed the privileges of the crossbowmen of the same city. These permissions allowed crossbowmen to carry their weapon with them about town, to wear the king's livery, and pardoned them should anyone be injured or killed because of the crossbowmen practising their shooting at butts – assuming those butts were set up with adequate safety precautions in place.[64] These efforts were carried on by his successors, Francis I (r.1515–47) and Henry II (r.1547–59). They both granted tax-exempt status to the winners of a variety of competitions for both crossbows and arquebuses, an early form of handheld firearm.[65]

Charles VII needed to reach beyond just the shooting fraternities to guarantee enough archers for his army. For example, the Paris crossbowmen had only sixty members and other fraternities would have been of a similar size or smaller. To solve this problem, he created a territorial militia called the *francs-archers*. This somewhat mirrored the efforts of Charles V to create a national militia of archers, as well as an earlier attempt on a smaller scale by the Duke of Brittany in the 1430s, but Charles VII had learned from his predecessors and was much more successful. In a 1448 ordnance Charles VII declared that in times of war each parish in the kingdom would be required to supply one archer or crossbowmen for every 120, 80 or 50 hearths in their parish, the number of hearths varying by region. The ordnance laid out clear instructions for the quality and type of equipment that the archer must bring with them. The men had to be of some standing, and the selection would be overseen by royal officials. The bowmen who were thus recruited benefited for their service by being exempt from all taxes, spared any obligation to provide billeting to any of the king's soldiers, and had any requirements for guard duty waived. It was these exemptions that gave the group their name: *francs-archer* means 'free archer'. These archers would also be paid 4 francs a month, which was the same pay as an archer in one of the lances in the *Compagnies d'Ordonnance*.[66]

The terms of the *francs-archers* were frequently amended over the following decades. In 1451 a new edict stated that each *franc-archer* was to be equipped at the expense of his community rather than his own, presumably to better ensure that the equipment standards set out in 1448 were met. There were also minor shifts in individual regional requirements as parishes haggled over how much service they ought to be obliged to provide. For example, in 1448 the city of Poitiers negotiated their quota of archers down from thirty to twelve, something they would do again in 1467 when they reduced their now-increased total of twenty-four back down to eighteen. In 1466 Charles doubled the size of the *francs-archers* to approximately 16,000, divided into four regions of 4,000 each, which required substantial renegotiation of the privileges and requirements for each region.[67]

Overall, it is hard to say what impact the *francs-archers* had on the end of the Hundred Years War. They were never tested in a pitched battle, but they must have served in some of the sieges that made up the end of the war. While the real star of those sieges was undoubtedly the gunpowder artillery train managed by the Bureau brothers, it still took a large force to successfully complete a siege and the *francs-archers* could have proved useful.[68]

At the end of the fifteenth century, with England defeated and torn apart by a civil war, the French monarchy turned its attention towards Italy and the Italian Wars that followed would substantially change both the French army and European warfare as a whole. Two major shifts happened to archers in both the *francs-archers* and the *Compagnies d'Ordonnance* during the Italian Wars. The first was a change in the ratio of the number of archers in the armies and the second was an alteration to what the term 'archer' meant in the context of a military unit.

As mentioned above, the standard ratio in French armies before Charles VII's changes was two men-at-arms to one archer. During the early years of the fifteenth century this shifted towards an even balance, even if on paper it should have had more archers, and in 1498 it was decreed that each lance would contain one man-at-arms and two archers. This was not a significant change from what Charles VII had ordered a half century earlier but the fact that it was reiterated at the end of the century may indicate that Charles' edicts had been aspirational rather than reflecting the composition of lances in practice. The size and number of lances fluctuated significantly throughout the reigns of Charles VII's successors and during a reorganization of pay in 1533–4 it was decided that a group of 100 lances would contain 100 men-at-arms and only 150 archers. Maintaining a permanent standing army was a huge cost for the French crown and not always one it was happy to pay, meaning that each monarch had to weigh the value of the army against what taxes he could raise and adjust the figures accordingly.[69]

The *francs-archers* had something of a more mixed time following Charles VII's death. Their value was hotly debated by successive monarchs and their performance was mixed. They performed well when campaigning in 1478 but disastrously in 1479. Even in 1477 their reliability had been being debated despite King Louis XI's (r.1461–83) demand that any *francs-archers* who were found to have abandoned their positions and gone home be punished as traitors. This all came to a culmination in 1481 when Louis XI abolished the *francs-archers* and ordered their equipment placed in central communal storage, which he then ordered to be inventoried in 1487. This was not the end of the *francs-archers* though, and they came back in one form or another from the 1490s, with no less an individual than Machiavelli commenting on them in 1510. In 1513 Louis XII (r.1498–1515) ordered a levy of 22,000 for frontier defence and in 1522 24,000 *francs-archers* were summoned to campaign in Picardy, Guyenne and Italy.

However, by this time a change had occurred in what was meant by an 'archer' in the French army. While previously the *francs-archers* had served with a bow or crossbow, the 1522 order decreed that archers must serve with a different sort of equipment. The change in types of armour is no great surprise, but now two-thirds of the *francs-archers* were to be armed with pikes while the final third would be a mix of halberdiers, crossbowmen and arquebusiers. All participants were to have swords or daggers as well. The *francs-archers* were levied in this form in 1523 and 1524 but local communities began to object that being forced to supply these troops was an unnecessary and undesirable burden. They were summoned one more time in 1525 before being abolished for good in 1535.[70]

This change to the meaning of a *francs-archer* was mirrored in the composition of the lances of the *gendarmerie*, as the *Compagnies d'Ordonnance* came to be known. By the end of the fifteenth century the 'archers' in a lance had begun to transform into lightly-armoured cavalrymen similar to men-at-arms. How and when this happened is hard to pin down, and it does not seem to have been a single sudden shift but rather a gradual back and forth over the decades. For example, in the 1490s the nobleman Robert de Balsac wrote with condemnation that many so called 'archers' were completely unable to shoot – showing that he expected the archers in a lance to have missile weapons. In 1515 and 1526 the king ordered that cities should maintain armourers to manufacture bows for the *ordonnances* and that captains should ensure the presence of 'a good number of archers and crossbow-men drawing the bow well from the saddle or on foot'.[71]

By 1549, however, the ordinance required the same armour of an archer as it did a man-at-arms, as well as a lance and a *pistolet*. This suggests that at least some difference still existed between the archer and the man-at-arms at this stage even if the former was no longer strictly a missile soldier. How exactly this transition happened is obscured by a lack of frank discussion of the shifts in archer's equipment in the surviving sources. Actions like the 1515 decree could have reflected an ideal rather than reality with the change already being well under way by that time.[72]

Despite these changes, including the wider adoption of handheld gunpowder weapons, there is clear evidence that traditional archery was present in French armies during the Italian Wars. The confusing terminology around archers can make pinpointing the exact contributions of bows and crossbows to the French war effort challenging. It is equally difficult to establish the last major engagement in which the French army used a significant number of crossbowmen. During his invasion in 1494 which started the whole conflict, Charles VIII (r.1483–98) led an army of 30,000 men which included a significant number of Gascon crossbowmen. The French general Gaston de Foix (1489–1512) was reportedly fond of the traditional crossbow and he brought roughly 2,000 Gascon crossbowmen with him into Italy in the early sixteenth century. These crossbowmen would have contributed to his famous victory at the Battle of Ravenna in 1512, but his death in that same battle probably cost the crossbow a valuable champion in the command of the French army. The crossbow was still hanging around at the Battle of Marignano in September 1515, but by this stage it seems to have been on the way out, largely replaced by the arquebus.[73]

By the Battle of Pavia in 1525 the crossbow had largely disappeared from French military tactics. After that battles became rarer in the Italian Wars which presents a further challenge to determining exactly when the crossbow disappeared from the French battlefield. Most historians have dated the end of the crossbow's time in the French army to sometime during the 1530s, but it is hard to be much more certain than that. Even after it left the battlefield the crossbow continued to be popular in France. As late as 1566 King Charles IX (r.1560–74) had to intervene via two separate ordinances to require the people of Paris give up their crossbows and bows and switch to using arquebuses instead. Clearly the crossbow continued to have its adherents long after it had been replaced on the battlefield by gunpowder weaponry.[74]

The crossbow in early modern Sweden

While the crossbow may have had its last hurrah on the Italian Peninsula sometime in the 1530s it continued to thrive elsewhere in Europe. The historian Josef Alm has argued that while the composite crossbow was generally the dominant ranged weapon in Scandinavian warfare, during the sixteenth century the steel crossbow began to replace it. This transition can be documented because sources helpfully distinguish between the two kinds of crossbows, calling them horn bows and steel bows, respectively.[75]

The Swedish author Olaus Magnus singled out the people of Östergötland as being particularly effective with the crossbow. He also claimed they had the most powerful crossbows, but it should be noted that Olaus himself was from Östergötland and may have been somewhat biased. Still, his point about the popularity of the crossbow in that region is not without merit. In 1551 a battalion from Östergötland led by Erik Mattson had 505 men equipped with steel crossbows compared to only 83 with arquebuses. That same year another battalion, this one led by a Nils Eskilsson, had 506 steel crossbows to 69 gunpowder weapons – a mixture of arquebuses and older hackbuts. But the men from Östergötland were not the only ones using crossbows this late. A Swedish battalion from Dalarna was equipped with 5 hackbuts, 361 arquebuses, 301 steel crossbows and 533 composite crossbows in 1554.[76]

This ratio of crossbows to gunpowder weapons was not to last, however, and during the Russian War of 1555–7 Swedish armies began to phase out the crossbow. The battalion from Dalarna had their 533 composite crossbows reduced to only three and their steel crossbows were similarly reduced. Most battalions were equipped with a mixture of polearms and gunpowder weapons, usually with a ratio of one polearm for every two gunpowder weapons. By 1560 in all the armouries in Sweden there were only 464 steel crossbows and 4 composite crossbows, a significant decline from only a decade before.[77]

While it became rarer the crossbow did not entirely disappear from early modern Swedish warfare. During the reigns of the Swedish Kings Erik XIV (r.1560–9) and John III (r.1569–92) the crossbow continued to be used in Swedish armies due to shortages in the number of available firearms. There is even a reference as late as 1644 of crossbows being carried by troops levied in Dalarna. While the crossbow was clearly no longer the preferred weapon in Sweden after 1555, it was still seen as an effective weapon and one that could be used when the king's first preference, firearms, were not available.[78]

The Great Siege of Malta

In 1522 the Ottoman sultan Suleiman 'the Magnificent' (r.1520–66) drove the Knights Hospitaller from the island of Rhodes, securing the eastern Mediterranean for his empire and removing a thorn that had long been in his side. The knights were homeless until 1530, when Emperor Charles V (King of Castile 1516–56, Holy Roman Emperor from 1519) offered them the island of Malta. Malta was no Rhodes, lacking the impressive fortifications built up over centuries and the abundance of natural wealth, but it had two excellent natural harbours and, most importantly, nobody was offering the Hospitallers anything better. Charles V required from them only a single hunting falcon, to be paid to the Viceroy of Sicily on Charles' behalf, along with a guarantee that the Hospitallers would never make war on his kingdom. The Hospitallers immediately set about trying to reinforce Malta and make it a new home from which they could conduct their ongoing war against the Ottomans. By this time the Hospitallers, once famous for their impressive castles in the Holy Land, were difficult to distinguish from pirates so it is no surprise that they immediately set up their new capital in the town of Birgu in Malta's main harbour. This left the main city on the island, Mdina, which was located further inland, in the hands of the Maltese.[79]

The Hospitallers had spent centuries building their fortifications in Rhodes, but they had neither the time nor the finances to achieve similar results in Malta before the next Ottoman attack. They also had limited local resources. Malta had ready supplies of stone, but it lacked much earth or wood – two materials necessary for fortress building. The soil was insufficient to grow grain in large quantities, so the Maltese imported much of their food from neighbouring Sicily and exported figs, melons and other fruit to pay for it. It also had few sources of natural water, with only a handful of springs located across the islands. However, this lack of abundance did offer one advantage to the Hospitallers: Malta was a difficult place to sustain a long-term siege, as any attacking army would be unable to live off the land. The island's rocky foundations also made any attempts to undermine the Hospitallers' walls and fortifications all but impossible.[80]

Over the next 35 years the Hospitallers constructed three main fortified positions around Malta's Grand Harbour. The first was the town of Birgu, built on a peninsula that emerged into the harbour perpendicular to the harbour's entrance. Birgu had originally been a fishing village which the order expanded into a respectably-sized town with walls and, on the peninsula's point, the fortress of Fort St Angelo. On a parallel peninsula located slightly further into the harbour they built the town of Senglea with the fort of St Michael located on the landward end of the peninsula. Between these two peninsulas a chain was strung that could be pulled up to block off access to the harbour that lay between. On the opposite shore from these two peninsulas was Mount Sciberras, a large peninsula that formed the north-western side of the Grand Harbour, with the smaller harbour of Marsamuscetto lying on its opposite side. Mount Sciberras was a problem because artillery placed on the mountain could fire upon the two towns. To prevent this, the Hospitallers constructed Fort St Elmo at the very tip of the mountain peninsula, overlooking the entrances to both the Grand Harbour and Marsamuscetto, allowing them to contest control of Mount Sciberras. For an image of the harbour approximately as it was during the siege, see Plate 29.[81]

Unlike at Rhodes, Suleiman, who at 70 still sat upon the Ottoman throne, did not lead the invasion of Malta in person. Instead, he chose as his general Mustafa Pasha (c.1500–80). Mustafa had participated in the siege of Rhodes as a young man, and Suleiman placed great trust in him. Commanding along with Mustapha was the admiral Piali Pasha (c.1515–78). Piali was married to one of Suleiman's granddaughters and thus had a greater claim to royal favour. The third commander of this expedition was one of the most feared men in the Mediterranean, Dragut (1485–1565), a notorious North African corsair and governor of Tripoli. Although Mustafa was given overall command, in practice these three men had to share in all the decision making. Suleiman sent to a truly staggering invasion force to Malta, estimated to be at least 30,000 soldiers, not including the sailors, slaves and other non-combatants that travelled with them. This included over 6,000 Janissaries, the elite of the Ottoman armies.[82]

Standing against the Ottoman invasion were the Knights Hospitaller and their Grand Master Jean de La Valette. La Valette had also fought at Rhodes as a young member of the order and had risen through the ranks over the intervening 40 years to reach its very top. He had filled numerous commands as a Hospitaller, and even spent some time in captivity as an Ottoman galley slave. Under La Valette the order had at most 700 knights in Malta out of a total of roughly a thousand. This force was supplemented by between 2,000 and 3,000 Maltese irregulars and a like number of mercenaries and Spanish soldiers supplied by King Philip II of Spain (r. 1556–98). The Hospitallers also had approximately 1,500 slaves – a mixture of captured Muslims and Christians prisoners – to use as a labour force to reinforce defences and do other onerous and thankless tasks. In total, the forces tasked with defending Malta probably numbered between 8,000 and 9,000 men, less than a third of the strength of the Ottomans.[83]

The Ottoman fleet arrived off Malta on 18 May 1565 and began to disembark the next day. At first sight of the Ottoman sails La Valette ordered all the available food and livestock behind the walls and distributed his forces between the handful of defensible positions on the island. Most were stationed in Birgu and Senglea. He placed his cavalry in Mdina, presumably hoping they could harass the Ottomans from the safety of that city. He placed smaller garrisons in St Elmo and the citadel on Gozo, a small island north of Malta. The Ottoman and Hospitaller armies first clashed on 21 May, when a small force sortied from Senglea and Birgu while the Ottomans were still deploying outside the cities. The battle lasted for six hours before La Valette ordered them to return. From then on, the Hospitallers fought a defensive war. Their plan was to buy time. On Sicily, 50 miles north of Malta, Don Garcia de Toledo (1514–77) was slowly amassing an army to drive off the Ottoman invasion and save the Hospitallers. Slowly was the key word and the Hospitallers and their allies would wait a long time for relief.[84]

In the meantime, after much debate the Ottomans decided on their battle plans. Mustafa had wanted to take Gozo, move on to Mdina, and then once the Hospitallers were completely isolated begin the siege of the Grand Harbour. Admiral Piali did not feel safe with his ships in their current harbour and wanted access to the bay north of Mount Sciberras. Securing that safe harbour would mean having to take Fort St Elmo first. In the end, Piali won the day, and the Ottoman forces started moving artillery onto Mount Sciberras to begin the bombardment and eventual assault on Fort St Elmo. The Ottoman chief engineer estimated that the fort would not last more than five days. On 24 May, after painstakingly hauling the guns into place, the Turkish bombardment of St Elmo began.[85]

The garrison of St Elmo was a mixture of Knights Hospitaller and allied soldiers numbering only a few hundred in total. They were now faced with the full attention of the main section of the Ottoman army. Mustafa ordered near constant bombardment of the fort, the Ottoman artillery slowly reducing its defences to ruins. The Ottomans also launched multiple assaults on the fort, only to be repulsed each time after hours of fighting. La Valette kept St Elmo supplied with a stream of new soldiers to replace casualties, but he had to balance defending the fort with the need to protect Birgu and Senglea. La Valette knew that most if not all the soldiers he sent over to St Elmo would die there, but he also needed them to hold out for as long as possible. Each week was thought to be St Elmo's last, with both sides doubting that it could hold out much longer, but somehow the defenders kept fighting.[86]

Finally, on 23 June St Elmo fell. The Ottomans killed every Knight they could. Nine Knights were reportedly taken prisoner, but no records indicate whether they were ever ransomed, while a handful of Maltese soldiers jumped into the Grand Harbour and successfully swam to Birgu, bringing with them the story of the final hours of the fort. St Elmo, which the Ottoman engineers had predicted would fall in five days, had held out a whole month. Historians have estimated that the Ottoman army lost upwards of 8,000 soldiers trying to take it. These casualties were significant, but the Knights had also suffered. The defenders lost at least 1,500 men in St Elmo, a significantly smaller figure than the Ottomans but one the Knights and their allies were less able to afford.[87]

The greatest price the Ottomans had paid was in time. They had expected to be finished with St Elmo in a week, but it was now near the end of June, and they had only just begun taking control of Malta. The passage of time presented two major risks to the Ottoman army. The first was that each passing week increased the likelihood that Don Garcia would organize his relief force and come to the Knights' aid. The greater risk, however, was the end of the campaigning season. Malta was not like Rhodes, where the Ottoman army had been only a few miles from the Turkish mainland. Mustafa and his army were halfway across the Mediterranean in unfriendly waters on an island with little food. They could not spend the winter there and even staying late in the autumn could risk them being unable to safely sail home through the stormy Mediterranean. The Ottoman army needed to conquer Malta before autumn, and it was now nearly July. With the clock ticking and casualties mounting – dysentery had now reared its ugly head in the Ottoman camps – Mustafa turned his attention to the joint siege of Birgu and Senglea.[88]

It took time to move all the guns from Mount Sciberras, as well as to dig trenches and expand the fortified camp outside of Birgu and Senglea. As a result, the bombardment of the two towns only began in earnest in early July. During this process 700 soldiers, mostly Spanish infantry but including a few more Hospitallers, landed in the north of Malta. On the night of 3 July, they sneaked around the Turkish camp and were ferried to Birgu, providing much-needed support to the beleaguered forces and in a single night replacing half of those lost in St Elmo.[89]

For his first attack on Birgu and Senglea Mustafa had something special planned. The Ottoman army dragged ships over the gap where Mount Sciberras joins the rest of Malta and into the bay on the western end of Senglea. The Hospitallers did their best to reinforce the exposed side of Senglea while under constant bombardment. On the morning of 15 July, the Ottomans launched a joint attack by both land and sea against both Birgu and Senglea. The fighting was fierce and bloody, but the attack stalled. In the end the Ottoman attack was repulsed with heavy losses. This failed attack set the tone for the following weeks.[90]

The Ottoman artillery kept up a near-constant barrage of the towns while Egyptian siege engineers attempted to undermine the walls. Occasionally Mustafa would order an all-out assault on the Hospitallers which, often through the vagaries of fate and to the surprise of both sides, would be driven back with significant Ottoman casualties. The attack on 2 August seemed to nearly be a complete victory when alarms sounded that the Ottoman camp was on fire and fears of the arrival of the Sicilian relief force caused Mustafa to pull back. In the end, it turned out to have been a surprise raid by the cavalry out of Mdina and not a massive Spanish army. The continuous bombardment and attacks were wearing the Christians down, but time was running out for the Ottomans.[91]

By the end of August, the defenders of Birgu and Senglea were exhausted. They had endured countless hours of cannon fire, their buildings were in ruins, and bodies littered the streets. On 29 August, it began to rain. The heavy rain stopped the Ottoman attack and limited both sides' ability to use firearms. Luckily for La Valette, the Knights had a large supply of crossbows in their armoury, a weapon whose performance was not at all diminished by the rain despite the claims of some fourteenth-century chroniclers. The crossbows were given out to the men defending the most dangerous positions and according to Francisco Balbi di Correggio (1505–89), a participant who wrote a very influential account of the siege, they were so effective that the Turkish soldiers quickly came to fear the Hospitaller crossbows. The crossbows were put to hard work against Ottoman attacks, which came steadily as the bad weather motivated Mustafa to try and seize the fortresses as soon as possible.[92]

On 7 September, the Spanish relief force finally arrived. Don Garcia's army numbered between 8,000 and 12,000 soldiers, nowhere near enough to drive off the Ottoman army when it first landed but enough to face the exhausted, diseased and depressed force that occupied the island at the end of the summer. With the impending arrival of autumn and the news of fresh troops landing on the island, Mustafa decided it was time to cut his losses and withdraw. After one final battle on 8 September, where Mustafa's exhausted forces were easily routed by the fresh and enthusiastic Spaniards, the Ottoman army retreated, leaving Malta devastated but still in the hands of the Hospitallers.[93]

The use of the crossbows in the wake of the rain in August is their most dramatic contribution to the siege, but it was almost certainly not the only time crossbows were used by the Hospitallers and their allies. In April, before the Ottomans arrived on the island, a small assortment of arms was issued from the Hospitallers' armoury to the city of Mdina. This included forty-one arquebuses and one crossbow. Given Mdina's peripheral role in the siege it is possible that this particular crossbow was never used, but it shows that the Hospitallers considered the crossbow a viable option for the coming battles, even if it was very much secondary to the arquebus. Stephen Spiteri has suggested that crossbows were primarily distributed to the Maltese irregulars. These made up a much larger proportion of the Christian defenders than the Knights, so if more of them were armed with crossbows then it is likely the crossbow was more widely used than the narrative of the siege would suggest.[94]

While the evidence suggests that the crossbows may have mostly been used by untrained militias, as recently as 1562 the crossbow was still standard issue on Hospitaller galleys and Knights had been expected to practice with the crossbow regularly. Nearly every Hospitaller at the Siege of Malta would likely have had some experience with the crossbow. The Palace Armoury in Malta still possesses a large collection of early sixteenth-century crossbows in the Spanish style, exactly what one would expect the Knights Hospitaller to have used at the time of the Great Siege.[95]

In Francisco Balbi's list of the soldiers who came to relieve the siege there is a single reference to an '*el capitan ballista*', a captain of the crossbows, in the retinue of the Marques Palavicino Rangone. If there was a captain of the crossbows, then there must have been several crossbowmen for him to captain. Whether these men were representative of the continued use of crossbows, or if they were an anomaly or possibly galley soldiers who had been recruited for land service is not clear from the single line reference in Balbi's diary. In any case, it seems likely that at least some crossbows were used in the final battle between the relief force and Mustafa in September of 1565.[96]

The End?

Was the Siege of Malta the last hurrah for the crossbow? As the previous sections have shown, the crossbow had fallen out of use in France in the 1530s and in Sweden in the 1550s, so how much longer after the 1560s could it have lasted? If the Siege of Malta was the siege of the century, then the Battle of Lepanto in 1571 was the battle of the century. A multi-national Christian fleet engaged in a bloody and brutal day-long battle with an Ottoman fleet, delivering a crushing defeat to the Ottomans and leaving thousands dead in the waters of the Mediterranean. It was a battle largely defined by arquebuses and cannons, as well as boarding actions and brutal hand-to-hand fighting. Amidst all this, so the story goes, Sebastiano Venier, the 75-year-old Venetian admiral, contributed to the fighting by standing on the deck of his ship and shooting at the enemy with his crossbow that his page reloaded for him.[97]

The crossbow ceased to be a common site on the battlefield sometime in the sixteenth century, but it did not entirely disappear. During the First World War some German soldiers looted two early modern target crossbows from their Belgian owners and reportedly used them to launch improvised percussion bombs. On the opposing side, a shortage of trench mortars caused several French soldiers to be equipped with crossbows, some with stirrups for spanning them, which could be used to shoot explosives. The French also offered a more modern take on the weapon in the form of the *sauterelle*, a man-sized crossbow designed to throw hand grenades into enemy trenches. In the mid-twentieth century the crossbow was considered by several militaries as a possible weapon for clandestine warfare since it is mostly silent and does not produce a muzzle flash. Elastic rubber was used in some experimental weapons, including a crossbow pistol intended for silent assassination. These rarely went beyond the experimental stage, but they show a continued fascination with crossbows as a technology and the potential that they could provide a valuable if niche service in modern warfare. The crossbow will probably never again be a common sight on a battlefield in Europe or anywhere else in the world, but it is far from a dead technology and militaries will inevitably continue to experiment with it.[98]

Chapter 7
The Sporting Crossbow

In 1223 King Valdemar II of Denmark (r.1202–41) was out hunting on the island of Lyø with his son and co-monarch Valdemar III (r.1215–31), alternatively known as Valdemar the Young to avoid confusion with a later Valdemar III (r.1326 – 30). While they were hunting the two of them were captured by Duke Henry I of Schwerin. The duke demanded that Valdemar surrender German lands he had conquered 20 years earlier and pay homage to the Holy Roman Emperor. Valdemar refused, hoping that his vassals would successfully pressure the German prince into freeing them. This continued for two years until forces loyal to Valdemar suffered a disastrous defeat in battle. His hope for rescue crushed, Valdemar was forced to give in to the duke's demands. While he would later have his vows to Duke Henry declared null and void by Pope Honorius III and resumed his wars in Germany, this must have been one of the worst hunting experiences in his life, that is until 1231.[1]

In November 1231, the Kings Valdemar once again went out hunting together, this time to Røsnæs, a peninsula on the north-west side of Zealand. The younger Valdemar was a recent widower, his wife having died in childbirth three months earlier, a tragedy that must have weighed heavily on both father and son. Possibly this trip was meant to get their minds off the recent tragedy, but it was to have its own tragic result. While out hunting on 28 November Valdemar III was accidentally shot by a crossbow and died. Subsequent myths and songs about the event said that he was shot in the heart – possibly by a fellow hunter named Eskild – and died immediately. There are reasons to doubt theses accounts: for example, the popular seventeenth-century folk song about Valdemar's death describes Eskild as using a steel crossbow, something was still at least a century away from being invented when Valdemar III was killed.[2]

Medieval accounts of Valdemar's death are less clear, with many not providing any details of his death. In the *Chronica Jutensis* his marriage to Eleanor of Portugal and both of their deaths are all reported within the span of a single sentence.[3] The brief account of his death in the *Vetus Chronica Sialandie* specifically says that he was shot in the foot which, if true, would suggest that much like Richard I he died of the infection from the wound rather than the wound itself.[4]

Later legends claim that Valdemar II ordered that as payment for the death of his son the entire forest on the peninsula of Røsnæs be destroyed, declaring that one should hardly be able to find a thorn left over from where the forest used to be. Valdemar III's body was eventually interred in the traditional site of kings of Denmark, Ringsted Abbey, where his wife was already buried. His father would join him there within a decade.[5]

An interesting aspect of the death of Valdemar III is that it was immediately accepted as an accident. This is a marked contrast with another famous king killed while hunting, William II of England (r.1087–1100). Also known as William Rufus, he was killed on 2 August 1100 while out hunting in the New Forest in Hampshire. Early accounts say that he was shot by an arrow by one of his own men, but later versions expanded that detail with all sorts of scandal. One popular telling identified Walter Tirel as the king's killer and claimed that the two were out hunting separate from the rest of the party. When Tirel killed the king, he abandoned his body, and it was later discovered by a local peasant. Several explanations arose about the death of William II. In one his death was divine retribution imposed upon an unjust or wicked king, the sort of thing that was technically an accident, but one made possible by God's guiding hand. In another King William was assassinated by rival nobles, possibly at the behest of his brother and successor Henry I – the same Henry who was almost killed by a crossbow wielded by his own daughter some years later – who was also a member of the larger hunting party on the day of William's death.[6]

While by most accounts William II was killed by an arrow shot from a bow, not a crossbow – although it bears noting that the earliest account does not specify the weapon and at least one historian has interpreted it as a crossbow– his death is an interesting contrast to that of Valdemar III. We see in one a tragedy mourned by all and in the other reports of scandal, divine punishment and possible assassination. Some of this difference can be explained by the possibility that Valdemar III was a better, or at least more liked, royal figure than William II. It did not help William II that he ruled a kingdom his father had only recently conquered, and that Norman internal politics had a reputation for backstabbing and infighting. William II was also succeeded by a brother – a natural figure for accusations of assassination – while Valdemar III had been co-monarch with his father, who continued to rule for several years after his son's death. This means that someone had something to gain from William's death while in Valdemar's case the kingdom lost an heir and its king a son. This would make the death of William ripe for salacious gossip but Valdemar III's an event to be generally mourned.[7]

The deaths of these two monarchs also serves as a reminder that medieval hunting could be extremely dangerous. These two are not even the only monarchs to die in an archery-related accident while hunting. Byzantine Emperor John II Komnenos (r.1118–43) accidentally cut himself on a poisoned arrow while on a boar-hunting expedition and died after it became infected when the emperor neglected to treat it.[8]

Medieval Hunting

Very little of the meat that European nobility ate came from their own hunting, instead they hunted primarily for their own enjoyment. In contrast, lower-status groups hunted for food or pelts which could either be used or sold on to others. The types of hunting also differed greatly within and between groups. Falconry and hunting with dogs, called venery, were very popular with the European elite but both were far beyond the financial means of most people. In contrast, hunting with bows and crossbows was far more equitable, with a wider range of social classes able to partake – although still with different overall goals for different classes.[9]

Unfortunately, despite its importance, medieval hunting has only received limited academic attention, and of that far more has been paid to the pageantry and practice of the elite hunting parties rather than to the more practical everyday hunters of medieval Europe. This is in part reflective of the surviving source material. Numerous medieval texts on hunting survive but these were all written by and for members of the nobility and so focused on the types of hunting which members of that class would participate in. This chapter cannot provide an overview of medieval hunting as a whole but will instead focus on several of the ways crossbows were used by hunters across medieval society.[10]

Hunting Crossbows

The crossbow has long been one of the most popular weapons for hunting, from the Middle Ages all the way to the modern era. The earliest depictions of crossbows in European art, the Gallo-Roman figures discussed in Chapter 3, are hunting scenes. The crossbow's ability to stay loaded made it ideal for stalking prey or lying in wait for the perfect shot. Its more compact shape made it easier to handle in dense forests than a longbow. The longbow was still a popular hunting weapon, especially due to its lower cost and the lack of a need for any spanning equipment to go with it, but the crossbow was the premier hunting weapon for those who could afford one.

At first crossbows that were used for hunting and those that were used for war were indistinguishable from each other. It is possible that in cases of elite hunting, such as those participated in by the unfortunate King Valdemar, the weapons may have had more decoration than those intended for war would, but for the more common hunter there was probably no difference between the weapon they would use to hunt and the one they would take with them to war.

With the advent of heavier bows, particularly with the invention of the composite crossbow, there may have begun to be some divergence in the design of hunting and war crossbows. The power of a bow required to kill deer or other large game would be much less than that of a weapon which

Stag hunt after a woodcut by Lucas Cranach the Elder, sixteenth century, Metropolitan Museum of Art, New York City. (Designated CC0 Public Domain)

would need to kill a fully armoured soldier. However, there are no surviving early composite crossbows, and the earliest surviving examples do not show any difference in design that would separate crossbows intended for hunting from those used in warfare. Composite crossbows must have been used for hunting but, as in warfare, they would not have entirely replaced wooden crossbows. Composite crossbows had the advantage of extra power over wooden bows, and they could also be made much shorter, allowing for a more compact weapon overall. On the other hand, composite crossbows took more effort to make and would be more expensive as a result. They also seem to have been heavier than wooden crossbows. This weight difference would not have been enormous, but it may have added up when a hunter was out for most of the day.[11]

With the invention of the steel bow the design of crossbows really began to diversify, which was probably the result of the development of specialized weapons for hunting and sport shooting. Steel bows could be made in a wider variety of weights, sizes and strengths than previous materials. Crossbow makers could create a small, light bow that could be easily carried for long distances in the woods, be spanned by hand, and that shot with enough force to kill prey. At the same time, however, steel could be used to make massive weapons of war with astounding draw weights able to threaten an armoured man-at-arms. The sixteenth century saw a staggering variety in the types of steel crossbows that were being made.[12]

Many late medieval and early modern steel crossbows survive in modern collections, giving a fascinating glimpse into the development of the weapon. The composite crossbow lingered on only into the first half of the sixteenth century, while in contrast the steel crossbow remained popular throughout the modern era. The diversity of the sixteenth century eventually gave way to a more standardized set of types of steel crossbow designed for hunting, target shooting or other regional sports.[13]

Gaston Phoebus

Gaston III, Count of Foix better known as Gaston Phoebus, was one of the most interesting characters in the history of the medieval hunt. Gaston lived an eventful and at times bizarre life. He stabbed his own son to death over the son's supposed attempt to poison him, a story vividly retold by the contemporary chronicler Jean Froissart (c.1337–c.1405). However, he is probably best known for his book, the *Livre de Chasse*. The 'Book of the Hunt' was a commentary on and description of contemporary hunting practices, in which Gaston was an enthusiastic participant. It was at least partially based on an earlier hunting book, the *Livre du Roy Modus et de la Royne Ratio* by Henri de Ferrières, but Gaston brought his own perspective to the work. At least forty-six copies of the *Livre de Chasse* have survived, a testament to its popularity, but the one that has drawn the most attention is the beautifully illuminated fifteenth-century copy known as MS. fr. 616 now held in the Bibliothèque nationale de France in Paris. This copy combined fabulous illustrations of medieval hunting scenes with Gaston's expert opinion on the subject to create the definitive late medieval hunting manual.[14]

Gaston divided his work into four main sections and an epilogue. These sections were: On Gentle and Wild Beasts, On the Nature and Care of Dogs, On Instructions for Hunting with Dogs, and On Hunting with Traps, Snares and Crossbow. The last section is the most relevant here, but it is worth considering how small a role the crossbow plays in the work as a whole. Crossbows were important enough to be mentioned in the section title but were still only a niche part of the types of hunting Gaston set out to describe.[15]

Gaston mentioned the use of crossbowmen and archers in his text in multiple places, often in their role in the pursuit of prey. He did not go into much detail about exactly what these archers were meant to do and provided little information on the specifics of the weapons they used. Nevertheless, the many mentions of crossbowmen show the importance of ranged weaponry to several of the styles of hunting Gaston was writing about. For example, while he primarily focused on the use of archery against larger game such as deer and boars, usually to drive them to where they could be trapped, he also included a description of the use of blunt-headed bolts to stun hares, which shows his understanding of the diverse uses of the hunting crossbow. The many depictions of bows and crossbows in the illuminations that accompany his text provide a valuable supplement to his sometimes-sparse descriptions. Some caution must be exercised since the art was made after Gaston's death, but the artists clearly worked from the text and the detail they add is worthy of examination in its own right. In the following pages a few of the more interesting combinations of text and art present in BnF Ms. Fr. 616 will be examined in further detail.[16]

Included among the folios are several scenes showing archers using crossbows to hunt. One of these images shows a crossbowman with several companions who are all using wooden longbows.[17] The crossbowman's weapon looks like it was made of the same material as the bows his fellow archers are using, which suggests that even in the fifteenth century wooden crossbows remained a popular weapon for hunting.

Another illumination shows four crossbowmen on the hunt – each is shown in a different stage of preparation with their weapon (Plate 30).[18] One of the crossbowmen still has his crossbow slung over his shoulder, the second is spanning his with a belt hook, the third is loading his bolt into place, and the fourth is taking aim at a nearby deer. Collectively they form a vignette of how a crossbow was used at the time.

Gaston's general description of the correct deployment of archers, while it only explicitly mentioned the longbow, would have been equally applicable to the crossbow. In MS. fr. 616, this passage is paired with an image showing archers with both bows and crossbows, so the medieval illuminators certainly believed this advice was equally applicable to both. It is worth quoting the passage at length:

> Placed at a stone throw's distance from each other, their backs against a tree, the archers, dressed in green for camouflage, await the game with bows strung, ready to fire off an arrow that has a sharp, barbed head. They know that they have to hit the animal's breast sideways on, at a shoulder, taking care never to aim in the other archers' direction. If the beast is only wounded, they whistle the dogs, slip them on its trail and the pursuit is on to the death.[19]

Gaston obviously shows concern about friendly fire on the hunt, warning archers to never aim in each other's direction. It is possible he was thinking of famous deaths like William II or Valdemar III when he wrote this, but more likely he had seen the risk of accidental shots in his many years of hunting – after all more than just kings were injured on medieval hunts. An example of the kind of camouflage he describes can be seen in an image with a mixture of crossbowmen and mounted hunters all dressed entirely in green, one even having a crown of vines on his head.[20] What is not clear in either this passage or in the associated illuminations is the extent to which Gaston thought that nobles would participate in the hunt by shooting bows or crossbows themselves or whether archers would generally be servants in a supporting role, like the hounds and their keepers.

One of the most intriguing illuminations in the BnF Ms. Fr. 616 accompanies a passage on using a cart to shoot game. The passage describes camouflaging a cart and locking its wheels to prevent it from making noise. One person would then drive the cart while another hid in the camouflage. The idea was to sneak up on unsuspecting game. The accompanying image shows a crossbowman adorned with leaves in the back of a simple cart as he and the driver – who is covered in leaves and mounted on a horse similarly adorned – sneak up on some unsuspecting beasts.[21] The crossbow itself appears to be either a standard wooden or composite crossbow and has a bolt with a barbed head loaded and ready to shoot. Both hunters are dressed entirely in green per Gaston's previous advice on equipping archers. It is hard to tell how common, or even practical, this method of hunting was but it is a fascinating view into the mind of a medieval hunter.[22]

Another page depicts an archer shooting at boars in a lake with his crossbow.[23] The archer is using barbed bolts and has a quiver with more. Unlike with the cart, the text Gaston wrote does not match the image at all. Gaston's text described how to find a place where boars leave their droppings and then advised erecting a platform near there where the archer could hide. He advised arriving before dawn and waiting for the boars so they could be shot while they were at ease. This method will be at least somewhat familiar to anyone who has ever hunted from a deer stand or duck blind – both modern methods of hiding oneself from prey. The image simply shows an archer walking up to the pond and taking aim at the boars – a very inadvisable strategy if the archer does not want to be attacked by an angry boar.[24]

When describing how to hunt a bear Gaston wrote:

> The bear's scent is looked for in fields, vineyards, oak and beech forests, with a good bloodhound on the leash. His strong odour, short gait, and droppings give the dog an easy task, and the groom has no trouble picking out the imprints of the hind paws. The hunt, however, is dangerous and only mastiffs can get near him as only they will run the risk of being knocked lifeless by one blow of his paw. Armed with bows, crossbows and pikes, the men follow the battle from horseback and try to put an end to the carnage as soon as possible with their weapons.[25]

The image shows two mounted men pursuing a bear in the woods accompanied by three men on foot – two with crossbows held over their shoulders and one with a spear – surrounded by a pack of dogs.[26] The passage and image together vividly convey a dangerous form of hunting. The image also shows how a crossbow might have been carried by medieval hunters while pursuing prey. The two archers hold their weapons differently; one positions the bow over his shoulder and holds the stock in his left hand while the other loops his right arm between the bow and the string and, tucking the weapon under his arm, has grabbed hold of the stirrup. Only the second archer has a clear spanning device – a belt hook hanging from his waist, a feature common to many crossbowmen in this manuscript. Both hold their bolts ready in their other hand, but since neither weapon is spanned, they are not exactly ready to shoot.[27]

Gaston's book was widely read–the number of surviving copies attests to that–but it was also translated and amended by later authors. The most famous version is the early fifteenth-century translation into English by Edward the Duke of York (c.1373–25 October 1415). Titled *The Master of the Game*, Edward translated parts of Gaston's original text while also incorporating new additions of his own, making a true hybrid work. Unfortunately for crossbow historians, Edward spent even less time discussing the role of archery in hunting than Gaston did so there is little to learn about the use of the crossbow from this revised work.[28]

The Crossbow and Aquatic Hunting

The crossbow has a long but somewhat overlooked history as a weapon for shooting aquatic prey. Sadly, while the crossbow was almost certainly used to hunt aquatic animals in the Middle Ages the best surviving evidence of its use dates from the modern era. One of the earliest references to the crossbow as aquatic hunting weapon is in a licence granted to Sir John Gate in 1547 by King Edward VI (r.1547–53). This licence permitted him to shoot, with either crossbow or gun, a variety of birds, otters, seals and all other fishes. At the time, all sea creatures were generally lumped together as fishes without distinguishing mammals from actual fish. This licence was granted under an act of King Henry VIII (r.1509–47) which had first introduced the requirement that hunters be licensed. It is very likely that in previous centuries hunters simply shot whatever fish they wanted and left no paper trail for historians to follow.[29]

Later records can give some insight into what kinds of creatures one might have hunted with a crossbow. In early modern France it was common to hunt frogs with a crossbow. In his painting *Members of the House of Orange Returning from Hunting*, the Flemish painter Sebastian Vrancx (1573–1647) included an archer using a specially-designed crossbow with a very long stock to shoot at frogs in a pond. Later frog-hunting crossbows closely resemble bullet cross-bows, many of them with barrels like a gun, rather than that shown by Vrancx. A reference in a 1789 issue of the *Gentleman's Magazine*, published in England, described the practice of shooting at pikes using a crossbow with a barbed bolt attached to a line. Another reference to this practice in 1801 suggests that this was primarily meant to kill errant pikes who preyed on other, more desirable, fish. The pikes themselves do not seem to have been eaten nor even hunted for sport. These were all small game in comparison to the most impressive target in all of crossbow hunting: whales.[30]

The crossbow was probably first used in whaling sometime in the sixteenth century but no evidence from that time has survived. In 1752 a Dr Bond mentioned in a paper delivered to the Royal Society that the crossbow had been used to throw harpoons but that it was too weak and was liable to break when used in the arctic waters off Greenland. In 1786 the *Transactions of the Society of Arts* posted a reward for anyone who could construct a crossbow for throwing harpoons at whales, but there is no evidence that anyone successfully claimed it. Despite these complaints, crossbows were used to hunt whales and several of those weapons have survived to the modern day. The purpose of these crossbows was not to kill the whale outright but rather to attach lines to it, forcing the whale to drag the whalers along after it. Once the whale was tired out, the whalers would close for the kill using a harpoon to spear it. While clearly a topic of discussion in eighteenth-century England, the crossbow seems to have disappeared from English whaling by the early nineteenth century.[31]

The crossbow's role in whaling lasted a bit longer in Scandinavia where it was popular for most of the nineteenth century. The weapon of choice for these hunts was a simple wooden crossbow that was spanned with a wooden version of a goats-foot lever. These weapons were probably not very powerful, but they could shoot a bolt with enough force to consistently penetrate the blubber of a small whale. Hunters would lure whales into an inlet or small fjord and trap them there by blocking the exit with a net. Then archers would either stand on the shore or in boats and shoot the trapped whales. After several shots the whales would become lethargic, at which point they could be finished off with spears and then pulled from the water for harvesting. It was not a very sporting way of hunting, but it was also small scale, and it is likely that no more than a handful of whales were killed this way each year.[32]

The crossbows used by these modern hunters show interesting design elements that make them distinct from earlier weapons. While the principles governing the making of the wooden bows were similar to those from the Middle Ages, the ones used for whaling were massive, far larger than the handful of surviving medieval examples. The goats-foot levers the whalers used were simple devices made only from wood, exactly the sort of device that medieval archers would have been able to construct, but as far as we know it has no precedent before the modern era. These crossbows and their levers are an excellent reminder that just because a technology was simple does not mean that it was old. These weapons were very much modern crossbows, they just used materials that were also available in previous centuries.[33]

Popinjay Shooting

Popinjay, also known as the *papegay* or 'parrot', was one of the most popular recreational shooting sports in Europe from the Middle Ages to the modern era. The Swedish author Olaus Magnus, writing in the sixteenth century, provided a good description of the sport that is worth quoting at length:

> However, for the most powerful crossbows the target consists of an artificial popinjay, which sits at the top of a rotating iron pole. This they seek to dislodge with blunt-headed bolts, so that it falls down. After the end of the competition those in charge are wont graciously to give the most skilful a prize, and at the same time to grant him some special right of priority, so that in future he can take his place among the highest ranking burghers.[34]

While Magnus was writing about the sport after the end of the Middle Ages, what he described was probably very similar to what the sport had looked like centuries earlier. Some aspects, such as the pole rotating, may have been a later addition but the core of the sport, a bird on a pole in a central place that archers shot at, was very old.

Popinjay shoot in Olaus Magnus, *Historia de Gentibus Septentrionalibus* (Rome, 1555).

Popinjay shooting probably originated in the thirteenth century and was well established in Europe by the fourteenth century. In 1354 the Grand Master of the Teutonic Knights ordered that a popinjay pole be constructed in every city in Prussia. Popinjay shooting was extremely popular in the cities of the Low Countries. Many Flemish cities hosted annual shooting competitions with huge numbers of participants. In 1454 and 1462 Bruges had 248 archers, their highest attendance, but their lowest number of participants, recorded in 1461, was still an impressive 169. These could also be major celebratory events, some Flemish towns allotted barrels of wine to the participating archers. These were not necessarily open to everyone, however, as some cities restricted participation in the popinjay to members of shooting guilds, institutions discussed later in this chapter.[35]

Popinjay shoots were major annual events for the cities and communities that organized them. They also presented an opportunity for potentially gaining prestige and advancing to the upper echelons of medieval urban society. On a wider scale they encouraged the practice of shooting – especially of the crossbow – which was a habit that was desirable to many urban communities and some local rulers. Popinjay shooting competitions were one of many ways that recreational sports could be used to promote familiarity with weapons of war among the populace – although how effectively the population could be tapped for military service varied greatly between regions and rulers. In the 1520s the city of Dijon was one of several to receive a royal confirmation of its annual '*papegault*' festival, with the decree praising how the sport trained the populace in the use of bow and crossbow for both recreation and the defence of the town. This was not a plan without risks for local rulers, however, as a populace trained in the art of war might decide not to do what they were told.[36]

There was room for variation in the specifics of the popinjay to suit local needs or preferences. Surviving records indicate that the pillars used in popinjay varied from 7m to 17m in height. The bird was usually brightly painted and about the size of a hen. Sometimes it came in multiple pieces, some of which could be shot off without dislodging the main bird. Prizes went to anyone who knocked off part of the bird and the grand prize went to whoever toppled the bird itself. The winner of the popinjay was usually given the title of shooting king and a badge to wear until next year's competition. In some regions if a single shooter won multiple years in a row, they might be given the title of emperor. Sometimes there were separate weapon-specific popinjay shoots for crossbows, bows and guns, but in other cases it was just a free-for-all of mixed weaponry.[37]

The most elaborate popinjay shooting competitions were held in the court of the Dukes of Saxony near Dresden. The dukes were very fond of the crossbow and the sport of shooting and held their annual competitions well into the nineteenth century. These were major events in the court of one of Germany's premier duchies and so came with all the ceremonial displays of wealth and power that necessitated. They were a far remove from the relatively modest origins of the sport.[38]

Crossbow Guilds

Crossbow shooting guilds were a fascinating overlap between the crossbow as recreation and the crossbow as a weapon of war. These were not guilds in the sense that most people would think of them – they were not organizations of merchants or artisans bound together with a set of rules to regulate their shared trade. Instead, they were something closer to sports clubs. Members of these guilds were generally well off and respectable men from urban centres who would gather to feast, socialize and participate in shooting tournaments. They were common throughout Europe with examples in France, Iberia, Switzerland, the Holy Roman Empire, the Low Countries and Italy. Genoa was so fond of the crossbow that when they banned tournaments for a time, they promoted crossbow shooting competitions as an alternative activity. A full study of crossbow guilds and the competitions they participated in could fill a series of books, so this section will focus on one of the best-documented regions: Flanders.[39]

The earliest evidence for Flemish shooting guilds is in the fourteenth century and they remained a powerful force until at least the sixteenth century. Shooting guilds were primarily urban organizations and were most common in areas with a large numbers of cities, like the Low Countries. They also had patron saints, St Sebastian for archery guilds and St George for crossbows. Large cities like Bruges and Ghent often had large guilds, and in some cases, they might even have more than one. Guilds frequently held large feasts which were a great opportunity for individuals to socialize and network both within their urban community and across the region. That is not to suggest that these were exclusively social clubs: members were expected to practice shooting their crossbows regularly. Members were given special dispensation that granted them a pardon in case they accidentally wounded or killed someone while practising, an important right to have in an urban environment. The guilds were generally required to maintain a shooting ground where members could practice. Sometimes these were near the guild house but often they were outside the city's walls.[40]

Several times a year the guilds would participate in shooting competitions. These events were an opportunity for guild members to wear their best uniforms and put on impressive displays of wealth and prestige, including arriving to the competition in elaborate flotillas or carriages. The terms of these competitions were published in advance, and announcements often included detailed descriptions of the specifications for the competition, such as the size of the bolts that would be permitted and the shooting distance. Competitions could vary from the popinjay shoots discussed above to a variety of competitions centred on shooting targets at fixed ranges. In general, popinjay shoots seem to have been more common for competitions between members of a given guild while inter-city competitions were usually team events where members of a range of guilds took turns shooting at targets in the main square of the hosting city. Competitions were huge affairs, sometimes lasting as long as three months and drawing hundreds of participants and thousands of onlookers. Winning these tournaments was a sign of great prestige for both the guild and the city they came from, and guilds took their participation very seriously.[41]

Guild members could also be drawn upon to use their weapons for more violent purposes. The guilds provided a core group of highly-trained crossbowmen that could be called upon to defend the city in times of war or to perform policing actions, sallying out from the city to drive off brigands or other sources of unrest. They could also be called upon to serve in war, although this varied

Men shooting at a target with crossbows, item no. 34350i, Wellcome Library, London. (Public Domain Mark).

greatly based on where they were located. The crossbow and archery guilds of the Duchy of Brabant, including the guilds in Brussels, had explicit privileges in their charters about their services owed to the duke as well as what the duke had to provide for them in terms of pay and equipment should they serve him in war. This relationship was put in place near the end of the fourteenth century and was a significant change from the previous system where each of the urban guilds had been required to provide soldiers to fight in the duke's wars. The quality of troops drawn from a mixed assortment of guilds was felt to be unreliable, so the duke opted to encourage the development of shooting guilds as an alternative source of better soldiers. This was a bit of a double-edged sword, however, as the duke's relationship with his cities was often fraught and by arming them with well-funded shooting guilds, he ran the risk of those guilds turning on him should the city rise in rebellion against their ruler.[42]

Many of the specific privileges of the Brabançon shooting guilds were wiped away when the Valois Dukes of Burgundy absorbed the duchy in 1430, but the new Valois rulers did not neglect the local shooting guilds. Duke John the Fearless (1371–1419) even participated in a competition while dressed as a guild member of the town of Oudenaarde, carrying his own weapon and reportedly winning two silver jugs. John's son Philip the Good (1396–1467) was a member of the Bruges and Ghent crossbow guilds. While he took part in urban guild events, he did so in his own livery rather than in the garb of an ordinary guild member like his father had. Both dukes made extensive use of crossbow guild members in their various wars, usually by hiring them as mercenaries.[43]

The crossbow guilds provide a valuable insight into the training regime of an elite group of crossbowmen. The promotion of shooting guilds by local rulers represented an alternative method for encouraging skilled archers as opposed to banning popular sports and mandating that the populace practice their shooting instead. The spirit of competition between the guilds helped to keep them well trained and ready to fight. This is not to say that the sole purpose for these guilds was to ensure a host of well-trained crossbowmen was available for local rulers. The guilds and their members had agency in directing their organizations and they were as much, if not more, the product of late medieval urban cultures than ducal policy. They show the balancing act that was maintained between the crossbow's use in war and its popularity in sport. The guilds endured long after the crossbow fell out of use in warfare, so one could say that the recreational aspect of the guilds ended up being the most enduring feature.

A Thoroughly Modern Weapon

The history of the crossbow in sport shooting did not end with the guilds of central Europe, nor with the eventual decline of the popinjay shoots of early modern Dresden. While its days as a target shooting weapon may be in the past – there is no Olympic crossbow shooting, after all – it continues to be a popular hunting weapon around the world. Most modern hunting crossbows, in the Western World at least, are of an entirely new type than those used in the Middle Ages. They are complex machines made from fibreglass with systems of pulleys and multiple layers of string that allow them to maximize their power while having a surprisingly low draw weight – at least in comparison to the most powerful medieval crossbows. These are highly efficient weapons with trigger systems similar to a modern rifle and often equipped with a scope or other sight enhancement to increase their overall accuracy. They are a clear sign that the crossbow never stopped advancing even as it was relegated to an ever-decreasing niche in the modern world.

Not all modern crossbow shooting uses these new types of weapons, however, and there continues to exist a thriving, if small, community of enthusiasts for the weapons of yesteryear. Constructing, shooting and even hunting with medieval and early modern crossbows continues to be a popular hobby in some parts of American and Europe. There are even books, such as David Watson's *Iolo's First Book of Crossbows*, that contain tips for constructing and maintaining a traditional crossbow for anyone who is interested. In Chapter 1 we briefly considered the work of Tod's Workshop which constructs and sells traditional crossbows to enthusiasts and hunters alike. While the crossbow's popularity as a hunting and sport weapon is much diminished, it is still hanging on and there is little reason to expect that it will vanish entirely.

Chapter 8

The Legacy of the Crossbow

William Tell

In the first decade of the fourteenth century three Swiss cantons which bordered Lake Lucerne – Uri, Unterwalden and Schwyz – were suffering under the yoke of oppression from their royal governors. The worst was Hermann Gessler, the viceroy of Unterwalden and Schwyz, who had been appointed by King Albert of Germany. In November of 1307 Gessler, in an act worthy of Caligula, had his hat mounted on a pillar in the town of Altdorf and ordered that everyone was to pay respects to the hat as if it were the viceroy himself. This, in addition to a policy of increased castle building and numerous petty acts of injustice, pushed the Swiss people to a point of crisis. A band of Swiss citizens from the three cantons met one night in an isolated meadow and took an oath, later known as the Oath of Rütli, that they would rise on New Year's and cast aside the rule of the imperial governors.

In the meantime, a local huntsman and his son were apprehended in Altdorf for failing to pay proper respects to the viceroy's hat. By coincidence, Viceroy Gessler arrived on the scene during the dispute between the huntsman and the viceroy's guards. Gessler knew the man was supposedly a great archer and, in response to a boast from the son, ordered that this huntsman should be forced to shoot an apple from on top of his son's head at 800 paces. Should the huntsman fail to take the shot, or should he miss, he would be executed.

The huntsmen drew two bolts and successfully shot the apple from his son's head with the first. When the viceroy inquired why he took out two bolts when he was preparing for his shot, he said that had he hit his son the second bolt was for Gessler's heart, and he would hit his mark. The huntsman was arrested for this insubordination and loaded onto a boat to be taken to be imprisoned in a fortress on the other side of Lake Lucerne. However, he escaped from the boat by leaping on to a rocky outcropping during a storm.

After escaping, the huntsman rushed off to lie in wait along the route that he knew Gessler would have to take when traveling back to his fortress. On that narrow path Gessler's bodyguard was separated from him by a nearby wedding procession, and while Gessler was distracted by a woman and her children begging him for mercy the huntsman shot the viceroy in the heart, ending his tyranny once and for all. Soon after Gessler's death the members of the Rütli tore

Das XV blat.
Von wilhelm Tellen dem frommen landt=
mañ der sinem eigen kind ein öpffel müst ab dem houpt schiessen
vnd wie es im ergieng.

[W]Un was ein redlicher mā im lande der hies
wilhelm Tell/der hat ouch heymlichen zů dem stöffacher vñ siner
geselschafft geschworen/der selbig gieng nun etwa dick vñ menig
C iiij

Print of the apple shot of William Tell from the *Chronicle of the Swiss Confederation* by Petermann Etterlin (Basel, 1507). (Public Domain Mark, Wikimedia Commons) down the Imperial castles, driving the other governors from Switzerland. By coincidence, King Albert was assassinated in early 1308, preventing any immediate response from the German monarchy to the Swiss uprising. Eventually a monarch more favourable to Swiss self-rule rose to the German throne, securing the rights that the huntsman, who by now you might recognize as William Tell, and the Rütli had secured in their uprising.[1]

The above is a much-abridged version of the story of William Tell as it was told in Friedrich Schiller's influential play *William Tell*, written in 1805. Schiller's play would go on to be a major inspiration for Gioachino Rossini's French-language opera *Guillaume Tell*, which included the famous William Tell Overture. Schiller's version of the William Tell story has been one of the most influential, but he did not invent the story out of whole cloth. Instead, he drew on medieval versions of the Tell story when writing his play.

Schiller's main source for his play was the *Chronicon Helveticum* by the Swiss historian Aegidius Tschudi (c.1532–72). This was a history of the Swiss from the year 1001 to the present, and throughout his text Tschudi incorporated copies of many valuable contemporary documents as well as providing a historical narrative. Tschudi's wrote his history as a manuscript, and it did not appear in print until 1734 when an abridged two-volume version covering the period 1001–1470 was published. It was this edition that Schiller used. More recently, the historian Bernhard Stettler edited a 22-volume critical edition and German translation of the *Chronicon Helveticum* published between 1970 and 2001.

The most significant difference between Schiller and Tschudi's version of William Tell is that Schiller combined two stories in his play: William Tell and the Oath of the Rütli. Both authors place the story of William Tell in the winter of 1307/08, but Tschudi placed the Oath of the Rütli in 1291. In Tschudi's version these were two separate stories that together made up the founding myths of Switzerland, but Schiller combined them into one shared story, setting it in 1307–08 since that allowed him to incorporate the actual historical event of King Albert's assassination in 1308, although the historical Albert was killed in May rather than in the late winter.[2]

While Tschudi's version of William Tell is arguably the most influential, it was far from the first. The earliest-known account of the story of William Tell was recorded in the *White Book of Sarnen*, a collection of medieval manuscripts compiled by the scribe Hans Schribe in or around 1474. The *White Book* was a compilation of earlier sources, so while it was assembled in the 1470s the texts were often much older. The version of Tell's story in the *White Book* does not differ much in its general detail from that described in the opening page of this chapter. While Tschudi and Schiller added more detail, the version in the *White Book* hits all the same story notes: the hat on the pole, the apple shot, Tell's leap from the ship, and his eventual assassination of the viceroy. The version in the *White Book* was not widely read during the Middle Ages and so probably had little immediate impact, but its longer-term influence was significant because one person who did read it was Aegidius Tschudi.[3]

The most famous element of the William Tell story, the apple shot, has an even older history. In the nineteenth century the historian Sabine Baring-Gould examined many different variations on the story that predate William Tell by centuries. In addition to stories of archers shooting apples off of loved ones' heads, there are also stories of even more bizarre medieval trick shots. One common alternative to an apple was a hazelnut, with one version changing the task to throwing a spear so that it passed between the hazelnut and the thrower's brother's head. Baring-Gould even found a story where the archer had to shoot a coin off someone's head. In his book Baring-Gould was particularly focused on whether William Tell was a historical figure or not, pointing to the many variations on Tell's most famous achievement as indicating the fictional nature of Tell and his story. Baring-Gould makes a strong argument for the fictional Tell but given that modern historians generally regard Tell as a folk hero rather than reality it can seem to modern readers like Baring-Gould was overly focused on just the apple shot and ignored the rest of the story. The story of William Tell, and his famous association with the crossbow, has a far wider reach than just a single feat of archery – Tell's role in killing the despotic ruler of his local community and the subsequent establishment of a semi-independent Swiss federation have had a wide impact in modern cultural memory, especially in his home country of Switzerland.[4]

One of the oldest and most influential versions of the story of William Tell is the *Tellenlied*, or the 'Song of Tell.' This song dates from the 1470s when the Swiss Confederacy was at war with the Valois Dukes of Burgundy. This song comes in several versions, but one popular form mentions the death of Charles the Bold (1433–77). The earliest surviving text of the song dates to 1501, but since many versions were likely in circulation simultaneously this should not be interpreted as a 'true' version of the song from which all others were derived. The version of the story in the song dwells primarily upon the apple shot, as many accounts of Tell have since, with a warning to tyrants wrapped up in the brief details around Tell's assassination of the viceroy and the driving of the governors from Switzerland. While not as complex a story as that found in the *White Book* or the *Chronicon Helveticum*, the *Tellenlied* would have been the most widely available version for the average individual in the Middle Ages and was arguably the most popular account of the deeds of William Tell before Schiller's play.[5]

The popularity of the *Tellenlied* also shows how the story of William Tell reverberated throughout Swiss history. The Burgundian Wars was another time when a neighbouring power, in this case the Valois Dukes of Burgundy, wanted to exercise control over Switzerland and its neighbours. The Swiss defeated the Burgundians at Grandson and Murten in 1476 but the culmination was the killing of Charles the Bold at Nancy in 1477 by a combined force of Swiss and German soldiers. These victories established even further Swiss independence from outside control while also popularizing the Swiss pikemen, who would go on to serve as mercenaries in many armies and conflicts over the following century. While it was with the pike and not the crossbow that the Swiss won these victories, the story of William Tell and his murder of a tyrant and driving off of foreign control clearly reverberated with the Swiss over a century and a half after the story was supposed to have originally taken place.[6]

The success of Schiller's play showed an enduring interest in the Tell story, and Rossini's opera only boosted that legacy further. In the same century, the newly-formed Swiss state took a serious look back on its origin myths, both Tell and the Rütli. In 1891 the Swiss government, only fully independent since 1848, decided to celebrate its 600th anniversary, and in doing so asserted the primacy of the Rütli story over that of Tell. While this decision was greeted warmly by some academics and government officials it did not go unchallenged and the average Swiss person on the street was far more likely to hold 1307, the year of Tell, in their mind as the origin of Switzerland. What would seem like a minor decision of when to celebrate a mythic origin of Switzerland sparked significant debate and no small amount of backlash at the time. People living in the modern era put meaning and national pride into the story of Tell and his crossbow shot, and in many cases, they still do. That shows both the power of stories in general as well as the importance of Tell and his crossbow to the Swiss people.[7]

The Legacy of the Crossbow

There was no definitive end to the crossbow, it being weapon with a long history that stretches to the present. Its full story is beyond the scope of a single volume, but this chapter will attempt to expand our understanding of the crossbow's legacy by examining two aspects that are easy to overlook: nationalism and colonialism.

The section about William Tell above has already briefly covered the importance of the weapon in the establishment of Swiss national identity. The next section will explore this issue from a different perspective by discussing the perennial debate about the crossbow versus the longbow and how that discussion is generally grounded in an English nationalist perspective of history. These two pieces have been paired to show one identity defined by the crossbow and another that has, at least in part, been defined in opposition to it.

The subsequent sections are case studies examining the use of European crossbows outside of Europe and the connection between the weapon and early modern colonialism. The first case study will examine the use of crossbows in the Spanish-Aztec wars and the European colonization of the Americas. The second will explore the use of the crossbow in early modern West Africa, including its possible introduction by European slavers.

The Crossbow versus the Longbow: a Study in Nationalism

The bow and the crossbow are overlapping technologies: they are both tools that shoot arrow-shaped projectiles with lethal effectiveness. While they are similar and both rely on the spring mechanics of the bow to generate their force, they have several important differences and these have sparked significant debate among historians and enthusiasts alike. At the same time, certain historical rulers and armies have shown a preference for one weapon over the other, most famously England's use of the longbow from the thirteenth to the sixteenth centuries. The reasons for preferring one weapon over another were probably numerous and may have included but not been limited to: cost, the quality of craftsmen available to make those weapons, difficulty in acquiring materials to make them, and the competency of the people who were inevitably recruited to fight medieval wars.

Historians and enthusiasts have long been drawn to a different explanation: the design of the weapons themselves. The theory goes that some difference in how these weapons functioned must explain why they were chosen for one army but not another. However, modern analysis of this topic comes with a lot of baggage. These debates are not just a matter of comparing two historical weapons but are instead tied up in matters of national identity and pride which greatly complicates the discussion. This is made even worse if these issues are overlooked or ignored.[8]

The most popular argument about the respective merits of the bow and the crossbow is the idea that the bow was the superior weapon in battle while the crossbow was better during sieges. A recent version of this argument can be found in David Green's *The Hundred Years War: A People's History*, as well as in Matthew Strickland and Robert Hardy's *The Great Warbow*, probably the single best book on the longbow. Medieval sieges have generally received less attention than battles so there is still plenty of room for historians to develop a better understanding of the subject, but overall, one thing that is clear is that crossbows were commonly used to great success in battles and longbows were often used during sieges.[9]

If crossbows really were better at sieges and bows superior in battles one would expect armies to have included a balance of both – after all sieges often caused battles and victories on the battlefield could lead to successful sieges. Edward III's victory at Crécy lead directly to the Siege of Calais and the famous Scottish victory at Bannockburn was the result of Edward II bringing an army to relieve the Scottish siege of Stirling Castle. These two types of warfare were not separate, they often flowed into each other, and a good commander needed to be prepared for both. It should be noted that it was quite common for medieval armies to include a mix of both crossbows and longbows, something these debates often overlook.

If they were similar weapons used in similar contexts, why have people spent so long comparing the bow and the crossbow? Part of it is that these kinds of discussions can be fun and are often harmless, but another important factor is the intertwining of English identity and the longbow. Since at least the time of the Tudors there has had a strong association with the longbow and the idea of the 'yeoman archer'. The common version of this belief creates a theoretical humble, common, 'yeoman' archer who served in the armies of England delivering crushing defeats to the elitist French knights, the flower of continental chivalry. This narrative was appealing because it presented the English archer as the common everyman proving his worth against impossible odds. It saw a significant resurgence of popularity in the wake of the First World War when many writers drew parallels between the 'yeoman' archer and the British 'Tommy' fighting in the trenches. This narrative overlooks that archers came from a range of backgrounds, as did the French soldiers they fought against. Both sides were commanded by elites and nobles, who had the most to gain from victory – and the least to lose from defeat.[10]

The idea of the humble English archer can be an implicit part of general works on English history or the history of archery, but it can also be more explicit in works. Take for example Warren Ellis' comic *Crécy* where the main character is a common archer whose manner and speech fit a stereotypical working-class Englishman. The comic's front cover is a hand making the rude two-fingered gesture that has mythically been associated with the English archer despite there being no historical basis for it. In this nationalist narrative the crossbow is the obvious opponent to the longbow, being the weapon associated with continental armies – despite English armies making ample use of it throughout the Middle Ages – and the Battle of Crécy offers the perfect story to cement all of this into an epoch of national mythmaking. Crécy is surpassed only by Agincourt which has the benefit of being memorably depicted by Shakespeare.[11]

This is not to say that there is no merit in discussing benefits of the crossbow and the longbow or trying to solve the question of why England used the longbow so much more than anyone else. However, these discussions are often so steeped in national myth and identity that they fail to resolve the questions they are meant to be answering. Consider as a point of comparison the level of emotional investment to be found in debating the respective merits of the pike versus the halberd. It is hard to imagine that debate summoning the same level of intensity. The degree to which the longbow has become the 'English weapon' has complicated discussions of both weapons and overshadowed the fact that the longbow was used throughout medieval Europe. Competing right next door to the crossbow guilds in Flanders were archery guilds specializing in shooting the longbow. The longbow was also used in medieval French armies, and Emperor Maximilian I of Germany showed a significant personal and military interest in the weapon. The longbow was a European weapon, not just an English one, and the crossbow was an English weapon as much as it was a European one.[12]

The Crossbow and Colonialism

Spanish conquistadors and the crossbow

It can be too easy to view Christopher Columbus' voyages west and the subsequent Spanish colonization of the Americas as clear markers of the end of the Middle Ages and the beginning of the modern era. History does not work that way though, and while there are aspects of the Spanish invasion that are distinct from, say, the Crusades there are many ways in which they were still the product of medieval Europe. This was true of their political institutions, worldview and how they waged war.[13] While the conquistadors brought with them the latest in gunpowder weapons, much of their equipment would not have been new to a European soldier from a century before. They brought swords, pikes, halberds and, most relevant to this book, many, many crossbows.

The collapse of the Aztec Triple Alliance in Mexico between 1519 and 1521 and the subsequent establishment of New Spain was arguably one of the most consequential wars in history. The alliance, which was dominated by the Mexica with the Acolhua and Tepanecs acting as lesser partners, lost their dominance and in its wake new power hierarchies arose. These included the new Spanish colonial project as well as a variety of native elites exercising political power in the region. This would have widespread repercussions across much of the world, but the greatest would be the depopulation of the Americas. Native populations were reduced by approximately 90 per cent from their pre-colonial levels by the actions of European colonial powers. In the Caribbean the exploitation and genocide of native populations pre-dated the fall of the Aztec Triple Alliance but the victory of the Spaniards and their allies enabled the expansion of this colonial project into even greater and more populous regions. This is also a highly contested subject, and historians of the period have to untangle a web of self-serving, one-sided and biased sources, some of which directly contradict each other, to even begin to understand what really happened. This is in addition to battling with centuries-worth of historiography written almost exclusively by European and other non-native historians viewing the subject through a European lens. Since the mid-twentieth century newer scholarship has increasingly challenged several fundamental assumptions about this period and pointed out many of the flaws in the traditional narrative of a brilliant Hernan Cortés (1485–1547) toppling an empire almost single-handed in only a few years.[14]

One of the most important correctives applied by these historians is to emphasize that the Spanish made up a minority of the armies that fought against the Triple Alliance. While many hundreds of Spaniards died in the war, the majority of combatants on both sides were natives. As an example, at the Siege of Tenochtitlan the total Spanish soldiers numbered less than a thousand but were accompanied by a force of over 25,000 Tlaxcaltec warriors. The Tlaxcalteca were subjects of the Aztec Triple Alliance, many of whom saw the Spaniards as an excellent opportunity to weaken the Mexica and their allies and place themselves in a new dominant political position. The Tlaxcalteca were not the only allies the Spanish had, they were also joined by the Huejotzinco, the Acolhua and the Chalca, among others. The story of a brilliant Cortés leading a handful of Spaniards to conquer a massive empire is a falsehood written as self-promotion by the victors, in many ways the Spanish-Mexica war was an internal Central American power struggle between different factions with the Spanish as just another player – and arguably one out of its depth at first.[15]

The story of one of the crossbowmen who followed Cortés and the myths that arose around him is useful example of this change in perspective. A classic moment in the Spanish conquest narrative is the meeting between Cortés and Emperor Montezuma (c.1466–1520) outside the city of Tenochtitlan. After the meeting, the Spanish were invited for an extended stay with the emperor. While in the city, the story goes, Cortés-captured the emperor and ruled the city through him as a puppet. Bernal Diaz (c.1492–1584), a conquistador who participated in the Cortés-led expedition but only wrote his account about it some 50 years later, told the story of a crossbowman named Pedro López who was one of the men assigned to guard the captive Montezuma. Diaz wrote that López insulted the emperor by cursing him and calling him a dog, swearing because he had to stand guard over him every night. The emperor was incensed and complained to Cortés who had the crossbowman whipped in punishment. López survived his whipping and even successfully escaped from the city when the inhabitants attacked the Spanish and drove them out despite surviving his poor treatment of the emperor.[16]

That is how the classic story goes, but in a recent book Matthew Restall argued that Montezuma was never actually a prisoner of Cortés and that this was a fiction concocted by the conquistadors after the fact. In fact, Montezuma ruled his city and may kept the conquistadors as amusements – conveniently housed next door to his personal zoo – or else hosted them due to obligations under native Nahua diplomatic conventions. In this version of events, López was not a guard of Montezuma the prisoner but rather used by an emperor interested in showing off his latest acquisitions and their fancy foreign weaponry. It is impossible to know the truth for certain, but Restall's version certainly sounds a lot more plausible than that recounted by Diaz or Cortés.[17]

According to Bernal Diaz, Cortés' initial expedition force consisted of 508 men, 109 sailors, and 16 horses. Among these men, Diaz singled out thirty-three crossbowmen and thirteen arquebusiers. This was not exactly an army fit for conquest, but it would be further reinforced before open hostilities broke out with the Mexica. Diaz puts the total size of the Spanish army before it suffered significant losses during *La Noche Triste* (30 June–1 July 1520), which was a fighting retreat from Tenochtitlan in the wake of Montezuma's death, at 1,300

Spanish soldiers, including 80 crossbowmen, 'a like number of musketeers', and at least 2,000 native Tlaxcaltec warriors.[18]

Up to two-thirds of the men initially mentioned by Diaz were likely dead in the wake of the *La Noche Triste* and many more crossbowmen would be sent to join the war over the following year, and even into the colonial period that followed. At the Siege of Tenochtitlan, which started in March 1521 and lasted for 75 days, the Spanish had been further reinforced and now included 118 crossbowmen, who were each given two strings for their weapons and could draw from a pool of over 50,000 crossbow bolts.[19] The detail about the two strings is reminiscent of an earlier event at the start of the expedition when Diaz described how Cortés ordered his men to check the status of their crossbows. This included verifying that the crossbows all had two or three spare nuts and strings each, as well as taking some test shots to ensure they were fully operational. One must wonder if this was a common practice in Spanish armies or if Cortés was uniquely particular about the state of his weapons.[20]

When describing the composition of the Spanish expeditionary force Diaz usually mentioned either more crossbowmen or an equal number of crossbowmen and arquebusiers. The historian Inga Clendinnen noted that in the terms agreed by members joining the expedition crossbowmen and arquebusiers were to be awarded equal shares of the spoils, a clear indication that Spanish commanders at the time saw them as equally valuable. This reflects a continuation of fifteenth-century tactics, when the crossbow and the then very new arquebus were frequently used side by side. Despite what one might expect, early gunpowder weapons did not replace crossbows and instead both weapons filled overlapping roles. Eventually gunpowder weapons won out–their greater overall power and the ability to pack soldiers armed with guns into tighter formations gave them a long-term advantage–but for the early history of the handgun they were complementary to the crossbow rather than a replacement.[21]

If only a few crossbows were used in the Spanish-Mexica war, then can it be said to have even meaningfully contributed to the war at all? Matthew Restall argued that the crossbow and arquebus were weapons that enabled the Spaniards to effectively establish defensive positions and drive off attacks from local warriors but did not provide them the tools for effectively launching a counter-attack. This would have been very valuable in the early stages of the Spanish expedition as they clashed with many different native communities. The use of the crossbow to defend against enemy attacks while in hostile territory is not significantly different from the tactics used by the Crusaders discussed earlier in this book. Diaz also described several times when Cortés sent crossbowmen and arquebusiers, always together, out on expeditions to search the surrounding country. However, it was arguably the steel sword rather than the gun or the crossbow that was the dominant weapon of the Spanish conquistador.[22]

In his description of a battle with some Tlaxcalteca in September 1519 Diaz described how the crossbowmen were instructed to work in teams, with some reloading the crossbows and others shooting them. This same technique was used at Jaffa in 1192 and this shows its enduring popularity throughout the centuries. It is likely that it was used during many other battles but went unmentioned by contemporary historians.[23]

The crossbow was not exclusive to the Spanish either. While in Tenochtitlan Cortés reportedly taught Montezuma how to shoot a crossbow, but this was more of a novelty. Of more relevance to the war were the native soldiers who forced several captured Spaniards to teach them how to use the crossbow. They put this newfound education to use by using their five captured crossbows to shoot at the Spaniards during the Siege of Tenochtitlan, although Diaz insisted that they inflicted no great injury upon the attackers.[24]

The fall of Tenochtitlan in 1521 did not mark the end of the end of the crossbow's use in Central and South America – it did not even mark the end of hostilities in central Mexico. Crossbows continued to be shipped over from Spain and its other colonies well after 1521 for use in suppressing the native populations and in further expeditions throughout the continent. The conquistador Pedro de Alvarado, who had participated in the campaigns against the Triple Alliance, took crossbows and arquebusiers with him in 1523 for his expedition overland toward modern day Guatemala, and he was not alone in this decision.[25]

Crossbows in the Americas

In 1518 Juan de Grijalva (c.1490–1527) led an expedition to explore the Yucatan peninsula, laying the foundations for where Cortés would go the following year. As part of his equipment, he brought several crossbows that were provided to him by Diego Velazquez (1465–1524), the governor of Cuba. The expedition charted the Yucatan coast but made no permanent settlement on the peninsula. In fact, several of his landing parties were driven off by the natives, suffering a number of casualties in the skirmishes. Subsequent voyages only stopped off in the Yucatan with no effort made at establishing a permanent Spanish colony until Montejo the Younger (1508–65) in the 1540s. Montejo also brought crossbows with him, showing that even 20 years after the fall of Tenochtitlan crossbows were a core part of the conquistador arsenal. The establishment of colonial rule in the Yucatan lacked the major pitched battles of central Mexico, but there was extensive guerrilla warfare as Spanish rule was limited to only part of the Yucatan until the eighteenth century, and it is likely that crossbows were used in that, as well as possibly fulfilling their other role as hunting weapons. That is not to say that the Yucatan was peaceful, as the Spanish tortured and executed countless natives, but these were usually done in practices closer to an inquisition than outright warfare and traditional weaponry played only a small part in these abuses.[26]

Francisco Pizarro (1478–1541) brought crossbows with him on his third, and most successful, expedition to Peru in 1530. The account of the expedition written by Pedro Sánchez de la Hoz (1514–47), who was Pizarro's secretary at the time, provides no detail on how many crossbows were brought and little information on how they were used. He frequently describes distances as being a 'crossbow-shot' but includes only one brief narrative description of their use during a campaign against Quizquiz (died 1535), one of the Inca Emperor Atahualpa's (r.1532–3) generals. Atahualpa had only recently triumphed over his brother in a civil war arising from the death of their father due to smallpox. It was in the wake of this unrest that Pizarro led his invasion, exploiting the divisions that were already tearing the empire apart. One can surmise from their presence in the campaign that crossbowmen played a part in Pizarro's war, but it is difficult to determine the extent of that role.[27]

Francisco de Orellana (1511–46) brought crossbows on his famous accidental expedition in 1541. Orellana was a lieutenant under Gonzalo Pizarro (1510– 48), the younger half-brother to the more famous Pizarro, who had been appointed as governor of Quito in what is now Ecuador in early 1541. In February of that same year Gonzalo led an expedition east to search for 'the Land of Cinnamon' – one of several quests for mythic places launched by conquistadors, the most famous being of course the quest for the golden city of El Dorado. The expedition reached the River Coca sometime after March and Gonzalo had Orellana lead an expedition down the river to find its source. Orellana and his men reached where the Coca joins the Napo River but when they travelled down the Napo, they realised that they could not travel back upstream against the current. After waiting to no avail for Gonzalo to follow they decided to continue down the river hoping it would lead them somewhere better. Orellana and his team travelled the entire length of the Amazon River, reaching the Atlantic Ocean on 26 August 1542, over a year and a half after they had left Quito. Orellana's team had three arquebuses and four or five crossbowmen with them, and they engaged in shooting battles with Amazonian natives on several occasions. Orellana's voyage brought the crossbow through the centre of the Amazon in the 1540s, showing how far the weapon reached geographically and chronologically.[28]

Friar Gaspar de Carjaval (c.1500–84) was among Orellana's party and wrote a fascinating description of the voyage, including many interesting anecdotes and details about the Amazon before the arrival of Europeans. He also included an amusing crossbow-related anecdote. Gaspar described how one of the Spanish crossbowmen decided to shoot at a bird while they were traveling down the river. When he pulled the trigger the crossbow nut flew out of the stock and landed in the river where it was promptly swallowed by a fish. Another Spaniard successfully caught the fish and they cut it open and retrieved the nut from its stomach, much to everyone's relief as, Gaspar recalled, 'it was the crossbows that saved our lives.'[29] This story captures the need to maintain the crossbows in good condition, Gaspar mentioned that they successfully repaired this crossbow, as well as the overall importance of the crossbows to the expedition's survival.

It would not be accurate to say that the crossbow was responsible for the successes of Spanish colonialism in the sixteenth century, but it is impossible to deny that it was a tool used in the establishment of those colonies. It is also worth taking a moment to contemplate the sheer scale of the violence that it contributed to as part of Spanish colonization. This book has covered many battles and sieges, moments of great violence that inflicted significant harm and death to many people, including innocents and non-combatants, but the scale of the genocide inflicted upon the native peoples of the Americas is unrivalled by anything in medieval European warfare. This extended beyond the violence of war or the diseases brought by Europeans. It was systematic and the result of deliberate practices like slavery and forced resettlement which killed many directly and made millions of others more vulnerable to diseases while preventing communities from recovering from the devastations caused by major epidemics. That the crossbow played some part in this genocide cannot be denied and must not be forgotten.[30]

The crossbow in West Africa

Much as the Spanish carried the crossbow across Central and South America, their Iberian neighbours the Portuguese brought crossbows with them on their ships sailing along the west coast of Africa. Vasco de Gama (c.1460s–1524) brought several crossbowmen with him on his famous voyage of 1497–9 which took him around the Cape of Good Hope and on to India. A contemporary journal of his trip mentions them being used during a series of engagements along the coast of modern-day South Africa.[31]

Half a century before de Gama's voyage Lançarote de Freitas led a pair of slaving raids on the West African coast. A roughly contemporary description by Gomes Eanes de Zurara of Lançarote's second raid, launched in either 1445 or 1446, mentions the use of crossbows by his men at several points. Lançarote and his companions participated in a dawn slave raid, with a force including crossbowmen and archers, on Tidra Island off the coast of Mauritania early in their expedition.[32] Sometime later crossbowmen under the command of Gomes Pires, one of Lançarote's captains, engaged in an archery duel with native Africans on Gorée Island off the coast of Senegal.[33]

During the same years of 1445 and 1446, Alvise Cadamosto (c.1432–88), a Venetian merchant in the service of the Portuguese prince, commanded two journeys along the Guinea coast. Cadamosto reported that the Africans feared the initial sound of the gunpowder weapons, but were unimpressed as the shots flew by, missing their targets, but the deadly accuracy of the crossbow proved a more lasting fear in their clashes.[34] Cadamosto also showed the weapons off to natives in what is now Senegal whom he had allowed to board his ship in the hopes of trading with them. He reported that they were much impressed with the weapon, which suggests that the crossbow was unknown in the region, or at least among people living along the coast.[35]

The presence of the crossbow on these raids should perhaps be no surprise, since unlike the Spanish wars nearly 70 years later personal firearms were still in their infancy in the mid-fifteenth century and crossbows were still the dominant ranged weapon at sea. While these stories show that the crossbow did not come to Africa peacefully, it did settle there, and the native Africans found their own uses for it. There is still a lot we do not know about the origins of the West African crossbow, including how widely it was used. There is very little archaeological evidence before the modern era and the written record is not much more robust. It has also not drawn very much scholarly attention, however, so there is still the possibility of uncovering more information through new research.[36]

The earliest written reference to the use of the crossbow by Africans is in the work of Leo Africanus (c.1494–1554). Originally a Muslim of Berber origins, Leo converted to Christianity after being captured by corsairs and handed over to the Knights Hospitaller. His most famous work is the *Description of Africa*, published in 1550. The book described several regions of Africa, based on Leo's own travels as well as with hearsay and reports he had picked up from other travellers. In his book he mentioned that slave traders traveling between Kano and Borno, now in northern Nigeria, would arm themselves with weapons, including crossbows, to drive off attacks by raiders and bandits.[37] It is possible that the crossbow had been adopted from the Portuguese in the previous century and slowly moved inland during Leo Africanus' lifetime. It is also possible that the crossbow was an established weapon in parts of Central Africa but remained unknown among the coastal communities who first encountered the Portuguese.

In 1512 King Alfonso I of Kongo (r.1509–42/3) recruited Portuguese mercenaries, including both musketeers and crossbowmen, to serve in his wars, although he was reportedly unimpressed with their performance. He also received a gift of two ornate crossbows from the King of Portugal that same year. This gift was significant as previously the Portuguese tried to prevent the selling of weapons – including guns, swords, daggers and crossbows – to Africans and they seem to have largely succeeded at this during the early periods of contact.[38]

Further evidence for crossbows in West Africa can be seen in several sixteenth-century Benin bronzes depicting Portuguese soldiers wielding crossbows. Two of these are currently in the British Museum – one is a bronze figure shooting a crossbow while the other is a relief image of a crossbowman holding his weapon in one hand and a bundle of bolts in the other. These bronzes show that in the sixteenth century the crossbow was still having an impact on African artists and communities, even if local rulers were more interested in adopting the still relatively new matchlock firearms for their wars rather than the older crossbow.[39]

'Though their specific equipment is not well detailed, an interesting example of the role of the crossbow African history can be seen in the use of archery by the Songhai who fought a 20-year guerrilla war against Moroccan invaders between 1591 and 1612. While the Moroccans' abundance of firearms brought them victories in pitched battles, which quickly caused the collapse of the ruling regime in the region, this changed when the resistance withdrew from the arid desert regions around Timbuktu to fight in the neighbouring swamps and mountains. In these environments gunpowder weaponry proved less effective against the guerrilla tactics of the Songhai. Moroccan fusiliers quickly learned to fear the threat of ambush and attack by the Songhai archers whose weapons still performed just as well in the damp swamps and were lighter and easier to carry over the mountains. Given that we know from Leo Africanus that crossbows were present in and around this part of central Africa nearly half a century before, it is plausible that the Songhai archers used a mixture of bows and crossbows in their resistance to the Moroccan mercenaries and their fusiliers. If this was the case this may mark the latest time when the crossbow played a major role in a war.[40]

It is necessary to clarify that no medieval or early modern African crossbows survive. Instead, historians have to draw conclusions based on later crossbows, mostly from the eighteenth or nineteenth centuries, that were brought to European collections as curiosities by explorers, merchants and slavers. Several traits are common across all known West African crossbows. They all had bows of wood, a long stock, and a peg-in-hole trigger. Some traits are common but not universal, such as the bow often being around 2ft in length and having the peg-in-hole trigger operated by a long lever that ran most of the length of the stock. Sometimes the stocks could be very long, two nineteenth-century examples now in the Deutsches Historische Museum having stocks over a metre in length, while others, such as a crossbow from Benin now in the British Museum, had stocks closer in length to a standard medieval European crossbow.[41]

One style of West African crossbow, probably native to Gabon, was very similar to the crossbows used by Norwegian whalers in the nineteenth century discussed in the previous chapter. The most immediately obvious difference between these weapons is the bows. The Norwegian weapons use massive yew bows, some over a metre long, while the African examples used a smaller bow, usually less than 500mm long. The other difference is that while the Norwegian weapons were generally spanned by a simple goats-foot lever, it seems that African examples were spanned by hand. The lack of a spanning tool could have necessitated the smaller bows, but just as likely the smaller bow could mean that no spanning tool was required. Both used very similar peg-in-hole trigger mechanisms, and both had long stocks, although this time it is the African crossbows that seem to have been longer.[42]

One other important difference between the Norwegian and African crossbows was the ammunition they shot. The Norwegian crossbows shot large bolts with iron heads. A similar iron-headed bolt was also used in West Africa for hunting large game, but the most common type of bolt was a dart made entirely out of wood, the tip of which was covered in poison. Hunters had to cut out the meat around the bolt when butchering their prey to avoid letting any of the poison mix in with their next dinner. The wooden darts were much smaller than the iron-headed bolts, but both types were shot from the same crossbows. Sometimes the darts were too small to easily rest on the crossbow and had to be stuck in place with gum. In the fifteenth century the Portuguese slave raider Alvaro Fernandez described the natives of Guinea as using poisoned arrows, a practice he could recognize as a form of poisoned bolt was still popular in Iberia in certain types of hunting. His testimony also shows that poisoned ammo predated Portuguese contact. This is not surprising when you consider that the poison would have been distilled from native plants or animals, meaning that learning how it was done in Iberia would have been of minimal use to African hunters.[43]

The similarity between the African and Norwegian crossbows raises some interesting questions about their origin. It is unlikely the Portuguese raiders of the fifteenth century were carrying massive yew whaling crossbows with them and it is also very unlikely that the weapon was introduced to West Africa by nineteenth-century Norwegian whalers. Henry Balfour, writing at the turn of the twentieth century, proposed that the Norwegian crossbow was more widespread in previous centuries and suggested that either Dutch or Danish ships could have introduced the weapon to West Africa in the sixteenth century. This is an interesting theory but is purely speculative. It is possible that both weapons could have a shared ancestor, with a similar crossbow having been introduced in both regions at around the same time. Without further research we are unlikely to solve how two similar weapons were adopted across such a wide geographic distance, and we may never know. It may remain yet another mystery in the history of the crossbow.[44]

Was the crossbow a European import or a West African original? The most widely-held theory is that the crossbow was introduced from Europe, and some African crossbows do suggest European influence, but we are drawing from a very narrow pool of evidence to make that conclusion. It is entirely plausible that the crossbow was invented in parts of Africa and then later African crossbow makers adopted aspects of European crossbow design. It is even possible that it came across the Sahara in the hands of Arab traders or crossed the Indian Ocean in the hands of Indian merchants and then slowly moved west over the continent. Regardless of the many possibilities for its first appearance in the region, it is clear that the Portuguese introduced the weapon to the West African coast at least. After that initial contact it proved a popular weapon in West Africa, used by hunters and warriors alike. It became an African weapon with its own distinct styles, not just a European import. There is plenty more we could learn about the African crossbow, and hopefully future research will provide a fuller explanation of its history while also uncovering even more mysteries for us to ponder.[45]

Conclusion

It can be too easy to take a dispassionate approach in examining the history of a weapon, to consider the weapon absent of its purpose, or at least without the gruesome details. It is therefore worth spending some time at the very end to consider the human toll of the crossbow. We have considered the deaths, wounding, and near misses of various famous medieval figures at the hands of the crossbow but these were but a tiny fraction of the damage wreaked by the crossbow in its prime. Even if we had accurate figures for medieval army size and war dead the body count of the crossbow would be beyond counting, to say nothing of the friends and families the dead left behind, that we do not have the data is no reason to ignore it.

We are not here to condemn a weapon from doing what it was designed and built to do, but we should also consider the impact it had and attempt to answer the question: what was it like to be faced with a crossbow in life-threatening circumstances? The famous French knight Geoffrey de Charny (c.1306–56) had the following to say on the topic:

> And you wouldn't want to miss an assault on a castle or tower. There you'll need to withstand some mighty blows, and fling yourself to the ground to avoid the shots of mangonels from the battlements. And quarrels, bolts and arrows fly about you – some of them are pinned to you and it's clear you're wounded. Now you're on the wall, where you're battered with blows so heavy that I'm not sure you'll make it back: everyone's out to get you! No one can save you now, I know, except God who ever watches over His own when He finds some good in them; to them He grants grace, rescuing them and sheltering them beneath His cloak.[1]

Geoffrey was not describing a specific event, rather providing a generic account of his own experience and in doing so gets to the heart of what it meant to be shot at by the crossbow and the despair that could come upon a soldier in life-threatening circumstances.

The anonymous author of the biography of Boucicaut (1366–1421), the marshal of France and a famous warrior, offered a slightly rosier version of the same type of situation:

> Then the attack began from all quarters, with the defenders displaying great vigour; the battle lasted until there were great numbers of dead and wounded on both sides. Châteaumorand found himself up against a stalwart defence on the seaboard, for the tower that guarded

> the harbour was well manned by archers and crossbowmen and men-at-arms who defended it with a hail of missiles. But had you been there, you might have seen our men, undeterred, use grappling irons and ladders to mount the walls, competing and vying to see who would be first on the rampart – but their ladders were too short, so they were unable to make the progress they would have wished.[2]

Again, we see the imagery of a hard-pressed attack under fire, imagery that continues to be popular in modern war movies. However, while Geoffrey attempted to place the reader in the shoes of the fighter in a general conflict, the anonymous biographer described a specific fight to praise his subject's great victory. Geoffrey was warning warriors of the risks of their profession and the anonymous author showed one of the clearest motives for why someone would put themselves in this situation in the first place: glory and reputation.

The two cases above both describe the assault in a siege, but what about other forms of warfare? In his *Collectanea* the master of arms Pietro Monte (1457–1509) had a few pieces of advice relating to the use of the crossbow in open warfare. His first piece of advice was: 'In provoking or skirmishing, when we are on foot, there is little danger as long as there are no crossbows, hand-guns, or bombards since we can easily evade or cover against any other projectile weapon . . .'[3] On the use of crossbows he advised: '. . . it is good to send in some troops who move lightly and force the others to come out, like crossbowmen and others who in our vernacular language are called *espingarderos* or *escopeteros* – troops using gunpowder weapons – since they can harass the enemies from a distance.'[4]

Neither of Monte's tactics were revolutions in terms of the use of the crossbow, but together they offer a more personal take on the weapon than can be found in general narratives of battles won and lost. Again, we see the reasonable fear of finding oneself opposite the crossbow, but Monte also includes the advantages of having crossbows on your side. These are two sides of the same coin, it is good to have the crossbow to draw your enemy to you for battle, but it is equally bad to find yourself opposed by the crossbow with an enemy attempting to draw you out.

Not everyone who faced a crossbow was a soldier, however, and many innocents found themselves on the receiving end of a raider's bolt. These stories are much harder for historians to access – dead peasants rarely left any written account and contemporary authors only occasionally concerned themselves with the details of the violence local populations were subjected to. The iconic late medieval images of St Sebastian suffering at the hands of archers and crossbowmen provide some small insight into what contemporaries believed it would have been like to be tortured and killed by these weapons. However, the commissioning and creation of these images was primarily the work of the upper classes in medieval society, so while they may reflect some aspect of the experience of the common people they are not of that community. Searching for their view may even be futile. After all, would the crossbow have stood out much to a group of peasants suffering from marauding and pillaging armies, or would it be just one of many weapons which inflicted great harm on them and their communities?

At the end of the day, no modern historian can place themselves in the shoes of a medieval soldier or civilian facing the consequences of medieval war. We can try our best to understand the mechanics of it, the overall strategy and the many tools that commanders and participants had at their disposal, but we cannot live their lives as they did. Contemporary accounts can give some insight into the experience, but it is no replacement for the real thing. This is something to be kept in mind when studying the crossbow or any weapon: these were tools for inflicting death and harm on human bodies in conflicts we can never fully comprehend. There was a human cost to their use that we must never forget even as we seek to better understand them.

Appendix: On Historiography

One of the most important parts of any historian's work is understanding what came before them. To badly paraphrase Newton paraphrasing Bernard of Clairvaux, all historians stand on the shoulders of giants, and it is important to give those scholars their due. The authors and works discussed below are all essential contributors to this book and their own works are some of the best places to start if you are interested in learning more.

One of the most influential and enduring historians of the crossbow is Sir Ralph Payne-Gallwey, whose book *The Crossbow* was first published in 1903 and is still one of the most widely available and accessible books on the subject. Payne-Gallwey established a broad narrative for the history of the crossbow, across two subjects: the extent of the crossbow's use in armies across Europe and the changes that crossbows underwent during the same time.

His book began with a short discussion of Roman siege weaponry, specifically the *ballista* and scorpion, because to Payne-Gallwey they were the predecessors of crossbows. However, the real narrative of the crossbow began when Payne-Gallwey reached the tenth century and crossbows first appeared in medieval textual sources. From there the evidence for the existence and use of the crossbow in Europe was overwhelming. England made quite substantial use of the crossbow up until the reign of Henry III when, Payne-Gallwey argued, the longbow replaced it nearly entirely. By the time of Henry V's Agincourt campaign, only 100 crossbowmen were to be found among the 30,000 soldiers sent to France. On the continent, however, the crossbow remained the dominant ranged weapon. The use of the crossbow in Europe only declined in the late fifteenth century with the introduction of reliable gunpowder weapons as an alternative.[1]

Payne-Gallwey's narrative for the development crossbow technology underwent was primarily linear. The oldest crossbows were simple and made of wood and would initially have been spanned by hand. Sometime during the Crusades, the composite crossbow was introduced to Europe, likely via Muslim craftsmen. To support this point Payne-Gallwey noted the presence of a 'Peter the Saracen' having been employed to make crossbows in England soon after the end of the Third Crusade.[2] The composite crossbow was the dominant form until the steel crossbow was invented sometime in the fourteenth century. The steel crossbow, in Payne-Gallwey's eyes, was the perfection of the weapon.[3] While his discussion of the wooden and composite crossbows was relatively brief, Payne-Gallwey devoted significant time to the examination of the steel crossbow, a weapon he believed eclipsed the famed English longbow in power.

Payne-Gallwey provided numerous line drawings and reproductions of crossbows and medieval images in his book. Many of them were from late medieval or early modern sources, which broadly reflects his greater interest and familiarity with the later period. He also included detailed examinations of crossbows in his personal collection, some of which he was still shooting at the time of writing despite their age, which makes him rather unique among historians of the crossbow. Many modern historians may have replica weapons, but few can claim to have a collection of centuries-old crossbows that they shoot.

Payne-Gallwey went into impressive detail on the various methods of spanning different types of crossbows. His description of spanning techniques connected different methods of spanning crossbows with different types of crossbow, linking the two technologies such that one type of spanning device was only used with one type of crossbow.[4] As far as Payne-Gallwey was concerned the more complex the spanning machinery the better, since this indicated that the weapon was more powerful.

Payne-Gallwey's book is the work upon which much of the field of crossbow studies rests, and for a book that is over 100 years old it holds up surprisingly well. It is not a book without its flaws, however. No one will be surprised to learn that significant historical research has been done in the past century and Payne-Gallwey very much reads like a scholar from another time with his sweeping statements and infrequent use of references.[5]

Arguably the best single history of the medieval crossbow was written in 1947 by Swedish historian Josef Alm. It was later translated into English and published in 1994 by the Royal Armouries.[6] In just over a hundred pages, Alm's book covered crossbows from the tenth to the sixteenth century with an emphasis on the latter two centuries. Despite being a much smaller work than Payne-Gallwey's, Alm's book went into much greater depth on the construction of crossbows. He spent almost no time discussing the general history of archery, including famous battles, or the longbow, and instead focused on smaller number of areas in greater detail.

Alm's examination of composite crossbows was much more thorough than Payne-Gallwey's and included a breakdown of the methods used to make composite bows for crossbows. His coverage of wooden crossbows, however, was severely limited by the lack of surviving examples.[7] Alm also included an in-depth discussion of how crossbow trigger mechanisms changed over the course of the Middle Ages, in part based on earlier research by Rudolf Wegeli.[8] For most of the text, however, Alm did not significantly disagree with Payne-Gallwey's general narrative of the history of the crossbow. Notably, however, he did not praise steel crossbows to nearly the extent that Payne-Gallwey did and instead argued that certain builds of composite materials could be just as powerful as a steel bow.[9] His narrative of the spanning devices used with crossbows agreed almost entirely in the order of their invention with Payne-Gallwey, although he did not agree that each device was paired with a certain type of crossbow.[10] Instead, Alm assigned spanning devices to crossbows based on the power of the crossbow rather than on their type.

Alm's book did not drastically alter the narrative history of medieval crossbows, but he greatly enhanced our understanding of the weapon, and his book is still one of the most detailed general works on the subject. If Alm's work had a major problem it is that he primarily had access to Swedish and German crossbows for his study, meaning that the work has something of a Central European bias.[11]

In 1990 the Society for Archer-Antiquaries published *A Guide to the Crossbow* based on research done by the influential archery historian W.F. Paterson. The book was published in memory of the recently-deceased Paterson, a prominent historian who wrote articles for the Society's journal, including several reports on the *Mary Rose* as it was being excavated.[12] His most famous work was *Saracen Archery*, a translation of a medieval Muslim manuscript on the practice of archery which he completed in cooperation with J.D. Lathan. *A Guide to the Crossbow* was a compilation of notes Paterson had written on the crossbow throughout the course of his life.[13] Since the book was a collection of notes it covers a wide range of subjects but does not provide a narrative history of the crossbow. Paterson's book made many significant contributions to the field in the areas he had studied. For example, he went into even more depth than Alm in his research of trigger mechanisms for crossbows. He engaged in relatively little discussion about the steel crossbow and instead most of his notes on bows were focused on wooden and composite crossbows. He included discussion of the Berkhamsted Bow, a surviving medieval wooden crossbow, as well as an examination of yew as it pertained to building crossbows. While Alm's book had a detailed account of composite bows, Paterson's included more cross sections as well as several surviving examples that Alm had not included in his book. While not a revolutionary work in the field, the individual sections of Paterson's book are all informative and represent a valuable contribution to the study of the crossbow.[14]

Two more recent English-language books on the crossbow are Dirk Breiding's *A Deadly Art: European Crossbows, 1250-1850*, which was published in early 2014, and Mike Loades' *The Crossbow*, published in 2018. Breiding was a curator at the Met in New York City and *A Deadly Art* was a showcase for the crossbows in the Met's collection rather than an overview of the weapon in general.[15] Each of the chapters begins with a few pages of introduction to a period in crossbow history before moving on to a detailed discussion of the contemporary crossbows in the Met collection. For example, the chapter on steel crossbows has five pages of introduction on the invention and spread of steel crossbows in Europe while the remaining thirty pages of the chapter contain detailed examinations of specific steel crossbows in the Met's collection.[16] Breiding included most but not all the crossbows owned by the Met, with a focus on the examples with beautiful artistic decorations.[17] Arguably Breiding's most valuable contribution to the discussion of late medieval crossbows was in his detailed study of Spanish crossbow design and its differences from Central European crossbow design. Spanish crossbows were a subject that Alm only briefly discussed and Breiding really emphasized how different these weapons were.[18]

Mike Loades' *The Crossbow* is a brief overview of the history of the crossbow with excellent illustrations. While not necessarily breaking much new ground it provides a valuable entry point for anyone interested in the crossbow. It is more accessible and up-to-date than Payne-Gallwey's classic work and lacks none of the visual impact of that older volume. It also contains a more robust discussion of ancient Chinese and Greek crossbow technology than any of the works discussed so far, excluding some sections of Paterson.

It may not be surprising to astute students of medieval weaponry that outside of the English-speaking community the crossbow is held in greater esteem. This has led to a greater supply of scholarship on the subject in other European languages, particularly German but also including French, Swedish and Polish, among others. This legacy can be traced back at least as far as Josef Alm's superb book already discussed above. However, unlike that book, many excellent works on the crossbow have yet to be translated into English and so have not penetrated English-language research as much as they should. A sample of standout non-English books will be discussed below. It should also be noted that while there is a much more robust scholarship in these languages than there is in English, the study of the crossbow remains a niche subject no matter what language you speak.

One book that is available in translation is Jean Liebel's *Springalds and Great Crossbows*, originally written in French and published in English as part of the same Royal Armouries series that included Alm's history. This is another short book, less than a hundred pages in total, that focuses very specifically on two easily-overlooked technologies. It should come as no surprise that Liebel's work is currently the definitive account of the great crossbow, a subject that was largely absent from earlier histories. Another excellent contribution to French scholarship is Valérie Serdon's *Armes du Diable*, which is a study of bows and crossbows in the Middle Ages. First published in 2005, Serdon's work is an excellent study of archaeological and artistic evidence, including one of the most robust examinations of medieval arrows and bolts currently available.

One of the most influential German books on the crossbow was Egon Harmuth's *Die Armbrust*. First published in 1975, it is a sort of translation and revised edition of Payne-Gallwey's *The Crossbow*. Harmuth made extensive updates and changes to Payne-Gallwey's book. He expanded the scope to include the early twentieth century and removed Payne-Gallwey's sections comparing the crossbow to the longbow and early firearms, as well as the long section on classical siege weaponry.[19] He also restructured the book into fewer chapters, each covering a more general topic. Payne Gallwey's work was divided into four sections with fifty-four sub-sections spread across those parts.[20] Harmuth excised the entire fourth section and its four sub-sections and re-divided his book into three parts with only thirty-six sub sections in total. While Payne-Gallwey's original used exclusively hand-drawn images, Harmuth added reproductions of medieval artwork and photographs of contemporary crossbows. Many, but not all, of Payne-Gallwey's original images were also retained in Harmuth's edition.[21] The most notable additions were images of the cross section of a composite crossbow, and several graphs which described the power of a steel crossbow.[22] The greatest single addition to the work overall was his examination of the mechanics of a steel crossbow.[23] Harmuth brought a much-needed update to a classic book and added several very important and new insights. It is unfortunate that his work is only available in German, as an English version would have made his updates to Payne-Gallwey more accessible to even more scholars.

Holger Richter's book *Die Hornbogenarmbrust*, published in 2006, is the most thorough and detailed study of the composite crossbow yet written and his *Die Kunst der Armbrustmacher in Dresden,* published in 2008, provides valuable insight into the often-overlooked role of crossbow makers. *Die Hornbogenarmbrust* includes twenty-eight surviving composite crossbows in its analysis of the technology. The illustrations and analysis include an unprecedented level of detail, with images and illustrations showing the internal structure of many composite crossbows' cores. To Richter, the composite crossbow represented a complex and sophisticated technology of war rather than a stopgap between wooden and steel crossbows. He supported the theory of an Islamic origin of the technology and argued that its eventual replacement by the steel crossbow in the sixteenth century as the result of the composite bows' complexity of manufacture rather than the latter weapon being a superior technology. The composite crossbow could not compete with the cheaper and easier-to-build steel crossbows.[24] The book was not limited to just archaeological analysis; Richter also discussed textual sources and wrote a general history of the composite crossbow.[25] Richter focused on the composite crossbow's design and the methods used to manufacture it. If there was a flaw in Richter's work, it was that he drew his archaeological evidence primarily from Central European museums, most of them based in Germany, Switzerland or Austria. However, most of the surviving composite crossbows are from this part of Europe so it is no surprise that it forms most of Richter's evidence and it could be argued that his book simply reflects the reality of where these weapons survive.

While he did not write a single, comprehensive history of the crossbow, Jens Sensfelder is one of the most important scholars of the crossbow currently working. Sensfelder has written many articles on aspects of crossbow history as well as editing the journal *Jahrblatt der Interessengemeinschaft Historische Armbrust*, but his greatest contributions have come from the detailed museum catalogues he has compiled. His most impressive work is his unpublished catalogue of Grandson Castle which fills three large binders. Each crossbow in Grandson Castle, of which there are over a hundred, was photographed and received at least two pages of description and discussion. While this work is Sensfelder's most impressive achievement, the fact that it was not published and can only be consulted by visiting the village of Grandson in western Switzerland has limited its impact.

Sensfelder's *Crossbows in the Royal Netherlands Army Museum* was a much more accessible work and was produced to the same rigorous standards as the Grandson catalogue. While the Grandson catalogue was entirely in German, *Crossbows in the Royal Netherlands Army Museum* was published in English, German and Dutch, making it accessible to a wider range of readers. The book included an introduction to the crossbow and a very general history which does not differ much from the accounts given in works already discussed here. What is very useful about this work is the level of detail Sensfelder put into his analysis. The crossbows he examined were measured in over a dozen different ways and he even included a discussion of the type of wood used in the stocks, something that few other authors mention.[26] The unique contribution that made this book stand out was the metallurgical analysis of the bow of a steel crossbow.[27] Performing this kind of analysis required the destruction of the bow so it is extremely rare for a collection to permit it, meaning Sensfelder's work is unique in having this type of study.

Most recently Sensfelder and Sven Lüken published *Die Armbrust Schrecken und Schönheit*, another excellent collection of articles on the crossbow with a detailed catalogue of crossbows in the Deutsche Historische Museum. In general, Sensfelder's catalogues contained far more detail than any other published work on the crossbow and his level of detail represents a standard that should be expected from all subsequent publications that include information on surviving crossbows.

There are of course many more historians whose contributions must be acknowledged. Polish historian Jan Kruczek's catalogue of crossbows held in Pszczyna Castle Museum and his history of the crossbow and crossbow makers, *Kusze I Ich Twórcy*, are both valuable contributions to the field in no small part thanks to the often-overlooked Polish crossbows he draws into the discussion. Oswald Graf Trap and Mario Scalini's catalogues of Churburg Castle are both very valuable for making one of Europe's largest and most remote collections of medieval weapons, especially crossbows, accessible to historians around the world. Lastly, credit must be given to Arthur Credland, Erhard

Franken-Stellamans, Sven Ekdahl and the many other historians who have written numerous articles and book chapters on the crossbow – many of which fill the Bibliography of this book. Every contribution to the study of the crossbow matters and if it was not for the many scholars who have worked, no matter how briefly, on this topic the book you are holding now would not exist. I can think of no more fitting ending than to give credit to what has come before me.

Notes

Introduction

1. John Gillingham, *Richard I* (New Haven, 1999), 321–34; Tony Hunt, *The Medieval Surgery* (Woodbridge, 1999), 26–9.
2. John Gillingham, 'The Unromantic Death of Richard I', *Speculum* 54;1 (1979), 21–3.
3. William Le Breton, 'Philippide de Guillaume Le Breton', in *Œuvres de Rigord et de Guillaume le Breton*, ed. Henri Delaborde, 2 vols (Paris, 1882–5), 2:52–3; 2:134–6.
4. Gillingham, 'The Unromantic Death of Richard I', 23–5.
5. Roger of Howden, *Chronica*, ed. William Stubbs, vol. 4 (London, 1871), 82–3.
6. Gillingham, 'The Unromantic Death of Richard I', 23–5.
7. Ibid., 28–31.
8. Ralph of Coggeshall, *Chronicon Anglicanum*, ed. J. Stevenson (London, 1875), 94–8; Gillingham, 'The Unromantic Death of Richard I', 26–8.8. Ralph of Coggeshall, *Chronicon Anglicanum*, 95–6.
9. Kelly DeVries and Michael Livingston (eds), *Medieval Warfare: A Reader* (Toronto, 2019), 72.
10. Ibid., 71–2.

Chapter 1

1. Anna Comnena, *The Alexiad*, trans. E.R.A. Sewter (London, 1969), 316–17.
2. Nicole Pétrin, 'Philological notes on the crossbow and related missile weapons', *Greek, Roman and Byzantine studies* 33;3 (1992), 282–3; Josef Alm, *European Crossbows: A Survey*, ed. and trans. G.M. Wilson and H. Bartlett Wells (1994, repr. Dorchester, 1998), 7–8.
3. Stuart Gorman, *Technological Development of the Bow and the Crossbow in Later Medieval Europe*, unpublished thesis (2016), 147–50; Robert C. Brown, 'Observations on the Berkhamsted Bow', *Journal of the Society of Archer-Antiquaries* 10 (1967), 12–17; Arthur G. Credland, 'Crossbow Remains (Part 2)', *Journal of the Society of Archer-Antiquaries* 24 (1991), 9.
4. Rudolf Wegeli, *Inventar der Waffensammlung des Bernischen Historischen Museums in Bern* (Bern, 1948), 22–5; Alm, *European Crossbows*, 15–18; W.F. Paterson, *A Guide to the Crossbow* (Oldland, 1990), 67–70; Holger Richter, *Die Hornbogenarmbrust* (Ludwigshafen, 2006), 46–55.
5. Gorman, *Technological Development of the Bow and the Crossbow*, 149–6, 166–9; Alm, *European Crossbows*, 15–19; Egon Harmuth, 'Concerning the One-Foot Crossbow of the High Gothic', *Journal of the Society of Archer-Antiquaries* 28 (1985), 9–11.
6. David S. Bachrach, 'Origins of the Crossbow Industry in England', *Journal of Medieval Military History* 2 (2004), 73–88; Alm, *European Crossbows*, 18–20; Paterson, *A Guide to the Crossbow*, 34–5; Ralph Payne-Gallwey, *The Crossbow: Its Military and Sporting History, Construction and Use* (London, 1903; repr. 1995), 62–3.
7. Payne-Gallwey, *The Crossbow*, 9.
8. Alm, *European Crossbows*, 34.
9. Bert S. Hall, *Weapons and Warfare in Renaissance Europe* (Baltimore, 1997), 17; Payne-Gallwey, *The Crossbow*, 6–9.
10. Dirk Breiding, *A Deadly Art: European Crossbows, 1250-1850* (New York, 2014), 30.
11. Paul E. Chevedden, 'Artillery in late antiquity - prelude to the Middle Ages', in *Medieval City under Siege*, ed. Ivy A Corfis and Michael Wolfe, (Woodbridge, 1999), 156–60.
12. Howard Blackmore, *Hunting Weapons: From the Middle Ages to the Twentieth Century* (London, 1971), 185–6

13. Alm, *European Crossbows*, 34–6; Payne-Gallwey, *The Crossbow*, 90–1; Sven Ekdahl, 'Horses and Crossbows: Two Important Warfare Advantages of the Teutonic Order in Prussia', in Helen Nicholson (ed.), *The Military Orders Volume 2: Welfare And Warfare* (Aldershot, 1998), 147–8.
14. David Grummitt, *The Calais Garrison: War and Military Service in England, 1436-1558* (Woodbridge, 2008), 120–1; Paul L. Holmer, 'The Military Crossbow in Yorkist England', *Journal of the Society of Archer-Antiquaries* 22 (1979), 11–14.
15. Alm, *European Crossbows: A Survey*, 19, 34; Paterson, *A Guide to the Crossbow*, 71; Payne-Gallwey, *The Crossbow*, 67–9, 101–07.
16. Breiding, *A Deadly Art*, 92–6; Gorman, *Technological Development of the Bow and the Crossbow*, 154–5.
17. Alm, *European Crossbows*, 33–4, 55–6; Arthur G. Credland, 'Crossbow Remains', *Journal of the Society of Archer-Antiquaries* 23 (1980), 12, 15–17; Payne-Gallwey, *The Crossbow*, 95–7; Jens Sensfelder, *Crossbows in the Royal Netherlands Army Museum* (Delft, 2007), 73–5.
18. Annette Holts Booth, 'Crossbow Production at the Archbishop's Palace, Trondheim, Norway', *Journal of the Society of Archer-Antiquaries* 39 (1996), 94–100; Gintautas Rackevicius, 'The Crossbow – The Weapon of the Invaders and The Defenders of Vilnius Castle (The Late 14th – Early 15th Centuries)', *Fasciculi Archaeologiae Historicae* 20 (2007), 64–6; Credland, 'Crossbow Remains', 12.
19. Alm, *European Crossbows*, 14–15, 33, 55–6; Paterson, *A Guide to the Crossbow*, 59–65.
20. Alm, *European Crossbows*, 14–15, 33, 55–6; Sensfelder, *Crossbows in the Royal Netherlands Army Museum*, 46–7
21. Alm, *European Crossbows*, 14–15, 33, 55–6; Payne-Gallwey, *The Crossbow*, 98–100; Harmuth, 'Concerning the One-Foot Crossbow of the High Gothic', 11.
22. Alm, *European Crossbow*, 9–10; Paterson, *A Guide to the Crossbow*, 16–18; Blackmore, *Hunting Weapons*, 179.
23. Payne-Gallwey, *The Crossbow*, 57–61; David S. Bachrach, 'Crossbows for the King: The Crossbow During the Reigns of John and Henry III of England', *Technology and Culture* 45;1 (2004), 109–13; Alm, *European Crossbows*, 15, 20–3, 34–5; Pétrin, 'Philological notes on the crossbow and related missile weapons', 280–91; Hall, *Weapons and Warfare in Renaissance Europe*, 17.
24. Jens Sensfelder, 'A Gothic Spanning-Belt Hook', *Journal of the Society of Archer-Antiquaries* 43 (2000), 27–8; Alm, *European Crossbows*, 20–3. Gorman, *Technological Development of the Bow and the Crossbow*, 192–3, 199, 206–08.
25. Alm, *European Crossbows*, 39–40; Payne-Gallwey, *The Crossbow*, 121–5.
26. David S. Bachrach, 'Crossbows for the King: The Crossbow During the Reigns of John and Henry III of England', 113–17; Breiding, *A Deadly Art*, 93; Paterson, *A Guide to the Crossbow*, 34–5, 49–50; Sensfelder, *Crossbows in the Royal Netherlands Army Museum*, 48–53.
27. Payne-Gallwey, *The Crossbow*, 140–4; Alm, *European Crossbows*, 38–40; Ekdahl, 'Horses and Crossbows', 143–7; Breiding, *A Deadly Art*, 94–5.
28. Breiding, *A Deadly Art*, 94–5; Gorman, *Technological Development of the Bow and the Crossbow*, 194–5.
29. Claude Blair, *European Armour circa 1066 to circa 1700* (London, 1958), 191–2; Gorman, *Technological Development of the Bow and the Crossbow*, 200–04.
30. Payne-Gallwey, *The Crossbow*, 121–5, 140–4; Alm, *European Crossbows*, 38–40.
31. Sensfelder, *Crossbows in the Royal Netherlands Army Museum*, 242–55; Gorman, *Technological Development of the Bow and the Crossbow*, 205–06.
32. Payne-Gallwey, *The Crossbow*, 81–3; Blackmore, *Hunting Weapons*, 188–9; Paterson, *A Guide to the Crossbow*, 44–5; Jean Liebel, *Springalds and Great Crossbows*, trans. Juliet Vale (Leeds, 1998), 29–32.
33. Alm, *European Crossbows*, 40–1; Arthur G. Credland, 'The Crossbow in the Far North', *Journal of the Society of Archer-Antiquaries* 26 (1983), 13–14; Payne-Gallwey, *The Crossbow*, 73–80; Breiding, *A Deadly Art*, 92–3.
34. Alm, *European Crossbows*, 27–9; Payne-Gallwey, *The Crossbow*, 84–91.
35. Breiding, *A Deadly Art*, 112–17; Gorman, *Technological Development of the Bow and the Crossbow*, 196, 204–05.
36. Alm, *European Crossbows*, 27–9.
37. Liebel, *Springalds and Great Crossbows*, 43–8.
38. Alm, *European Crossbows*, 42–4; Breiding, *A Deadly Art*, 118–20; Carsten Rau, *European Arrowheads and Crossbow Bolts* (Berlin, 2018), 185–215; Valerie Serdon, *Armes Du Diable: Arcs et Arbaletes Au Moyen Age* (Rennes, 2005), 87–126.

39. Sensfelder, *Crossbows in the Royal Netherlands Army Museum*, 274–7.
40. Josef Alm, *European Crossbows*, 42–4; Breiding, *A Deadly Art*, 121–2; Sensfelder, *Crossbows in the Royal Netherlands Army Museum*, 280–1.
41. Alm, *European Crossbows*, 42–4; Breiding, *A Deadly Art*, 122–3; Blackmore, *Hunting Weapons*, 193–7; Rau, *European Arrowheads and Crossbow Bolts*, 216–26.
42. Mark Denny, *Their Arrows Will Darken the Sun: A Guide to the Evolution and Science of Ballistics* (Baltimore, 2011), 7–15; Erhard Franken-Stellamans, 'Der Bolzenflug', *Jahrblatt der Interessengemeinschaft Historische Armbrust* 5 (2013), 64–78.
43. Tim Baker, 'Bow Design and Performance', in *The Traditional Bowyers Bible*, ed. Jim Hamm, 4 vols (Guildford, 1992), 1:96–103.
44. Ibid., 46–74; Mark Denny, 'Bow and Catapult Internal Dynamics', *European Journal of Physics* 24 (2003), 367–72, 377–82; Jim Hamm, 'Tillering', in *The Traditional Bowyers Bible*, ed. Jim Hamm, 4 vols (Guildford, 1992), 1:257–86.
45. Baker, 'Bow Design and Performance', 68–73.
46. Jim Hamm, 'Sinew Backing', in *The Traditional Bowyers Bible*, ed. Jim Hamm, 4 vols (Guildford, 1992), 1:213–4, 229–32.
47. Denny, 'Bow and Catapult Internal Dynamics', 377–8: *Arab Archery*, trans. Nabih Amin Faris and Robert Potter Elmer (Princeton, 1945), 10–16; Adam Karpowicz, 'Ottoman Bows – An Assessment of Draw Weight, Performance and Tactical Use', *Antiquity* 81 (2007), 675–83; Paul E. Klopsteg, *Turkish Archery and the Composite Bow* (London, 1934), 30–40; Brian Cotterell and Johan Kamminga, *Mechanics of Pre-Industrial Design* (Cambridge, 1990), 185–7.
48. Egon Harmuth, *Die Armbrust* (Graz, 1975), 125–36; Erhard Franken-Stellamans, 'A Mathematical Method for Determination of the Appropriate Draw Length for a Given Steel Bow', *Journal of the Society of Archer-Antiquaries* 47 (2004), 92–7.
49. Ekdahl, 'Horses and Crossbows', 147–8; Sven Ekdahl, 'Die Armbrust in Deutschordensland Preussen zu Beggin des 15. Jahrhunderts', *Fasciculi Archaeologiae Historicae* 5 (1992), 22–6, 45; Credland, 'The Crossbow in the Far North', 13–14; Sensfelder, *Crossbows in the Royal Netherlands Army Museum*, 361–7; Blackmore, *Hunting Weapons*, 186.
50. Robert Hardy, J. Levy, P. Pratt, Roy King, S. Stanley, B.W. Kooi, D. Adams and A. Crowley, 'Assessment of the Working Capabilities of the Longbows', in *Weapons of Warre: The Armaments of the Mary Rose,* ed. Alexandra Hildred, 2 vols (Portsmouth, 2011), 2:627.
51. Denny, 'Bow and Catapult Internal Dynamics', 372–6; Cotterell and Kamminga, *Mechanics of Pre-Industrial Design*, 181–5; Baker, 'Bow Design and Performance', 44–9, 69–74.
52. Paterson, *A Guide to the Crossbow*, 31–2.
53. Ibid., 30–1
54. Alan Williams, *The Knight and the Blast Furnace: A History of the Metallurgy of Armour in the Middle Ages & the Early Modern Period* (Leiden, 2003), 927–49.
55. Liebel, *Springalds and Great Crossbows*, 61–8; Williams, *The Knight and the Blast Furnace*, 927–49.
56. Gorman, *Technological Development of the Bow and the Crossbow*, 184–5; Brown, 'Observations on the Berkhamsted Bow', 12–17; Paterson, *A Guide to the Crossbow*, 31–2.
57. Franken-Stellamans, 'A Mathematical Method for Determination of the Appropriate Draw Length for a Given Steel Bow', 92–4; Harmuth, *Die Armbrust*, 125–36.
58. Alm, *European Crossbows*, 33–4, 55–6; Payne-Gallwey, *The Crossbow,* 95–7; Sensfelder, *Crossbows in the Royal Netherlands Army Museum*, 73–5.
59. Thom Richardson, Ballistic Testing of Historical Weapons', *Royal Armouries Yearbook* 3 (1998), 51.
60. Keith Watson, 'Estimating Longbow Weights from the Arrow Shaft Data', in *Weapons of Warre: The Armaments of the Mary Rose*, ed. Alexandra Hildred, 2 vols (Portsmouth, 2011), 2:617–18; Richardson, 'Ballistic testing of historical weapons', 52; Brown, 'Observations on the Berkhamsted Bow', 12–17.

Chapter 2

1. William of Poitiers, 'Gesta Willelmi', in *The Battle of Hastings: Sources and Interpretations*, ed. and trans. Stephen Morillo (Woodbridge, 1999), 3; 15.
2. Ibid., 14; Colm Flynn, *Crusader Artillery, 1097-1148: Typology, Terminology and Character*, unpublished thesis (2011), 101.

3. 'Carmen de Hastingae Proelio', in *The Battle of Hastings: Sources and Interpretations*, ed. and trans. Stephen Morillo (Woodbridge, 1999), 45–6, 50; Jim Bradbury, *The Battle of Hastings* (n.p. 1998; repr. 2010), 115–17.
4. William of Jumièges, 'Gesta Normannorum Ducum', in *The Battle of Hastings: Sources and Interpretations*, ed. and trans. Stephen Morillo (Woodbridge, 1999), 18.
5. 'Anglo-Saxon Chronicle', in *The Battle of Hastings: Sources and Interpretations*, ed. and trans. Stephen Morillo (Woodbridge, 1999), 24–5.
6. Bradbury, *The Battle of Hastings*, 114–15; Lewis Thorpe, *The Bayeux Tapestry and the Norman Invasion* (London, 1973), 24–31.
7. Bradbury, *The Battle of Hastings*, 149–51; David Bernstein, 'The Blinding of Harold and the Meaning of the Bayeux Tapestry', *Anglo-Norman Studies* 5 (1982), 40–64.
8. Martin K. Foys, 'Pulling the Arrow Out: The Legend of Harold's Death and the Bayeux Tapestry', in Martin K. Foys, Karen Eileen Overby and Dan Terlka (eds), *The Bayeux Tapestry: New Interpretations* (Woodbridge, 2009), 168–79.
9. William of Malmesbury, *Chronicle of the Kings of England*, trans. J.A. Giles (London, 1847), 277–8; Bradbury, *The Battle of Hastings*, 118–19; Foys, 'Pulling the Arrow Out: The Legend of Harold's Death and the Bayeux Tapestry', 164.
10. Henry of Huntingdon, *The Chronicle of Henry of Huntingdon*, ed. and trans. Thomas Forrester (London, 1853), 212; Bradbury, *The Battle of Hastings*, 119.
11. *Domesday Book* https://opendomesday.org/place/SP2050/ailstone/
12. Blackmore, *Hunting Weapons*, 175; See also: https://opendomesday.org/place/SO7619/bulley/https://opendomesday.org/place/TG4704/burgh-castle/
13. Nicholas James Crosby Smith, *Servicium Debitum and Scutage in Twelfth Century England With Comparisons to the Regno of Southern Italy*, unpublished thesis (2010), 150–2.
14. David S. Bachrach, 'Crossbows for the King: The Crossbow During the Reigns of John and Henry III of England', 109–11, 118–19; Harmuth, 'Concerning the One-Foot Crossbow of the High Gothic', 9–11; Alm, *European Crossbows*, 19–24.
15. David S. Bachrach, 'Crossbows for the King: The Crossbow During the Reigns of John and Henry III of England', 113–17; David S. Bachrach, 'Crossbows for the King, Part II: The Crossbow During the Reign of Edward I of England (1272-1307)', *Technology and Culture* 47;1 (2006), 83–5; Payne-Gallwey, *The Crossbow*, 121–5; Credland, 'Crossbow Remains (Part 2)', 11–12.
16. Guy Wilson, 'What's in a Name? One-foot and Two-Foot Crossbows', in *ICOMAM 50: Papers on Arms and Military History 1957–2007*, ed. Robert Douglas Smith (Leeds, 2007), 300–25.
17. David S. Bachrach, 'Crossbows for the King: The Crossbow During the Reigns of John and Henry III of England', 111–13; Credland, 'Crossbow Remains (Part 2)', 9; Payne-Gallwey, *The Crossbow*, 58–9; Wilson, 'What's in a Name? One-foot and Two-Foot Crossbows', 303–14.
18. Guy Wilson, 'What's in a Name? One-foot and Two-Foot Crossbows', 314–18; Thom Richardson, 'Springald Sizes in the 14th Century', in Robert Douglas Smith (ed.), *ICOMAM 50: Papers on Arms and Military History 1957–2007* (Leeds, 2007), 327–8; Serdon, *Armes Du Diable*, 87–126.
19. Paterson, *A Guide to the Crossbow*, 31–2.
20. Wilson, 'What's in a Name? One-foot and Two-Foot Crossbows', 318–24; Gorman, *Technological Development of the Bow and the Crossbow*, 147–9, 150–3.
21. Andrzej Nowakowski, *Arms and Armour in the Medieval Teutonic Order's State in Prussia*, trans. Maria Abramowicz (Łódź, 1994), 98–9.
22. Breiding, *A Deadly Art*, 30–4, 47–50; Paterson, *A Guide to the Crossbow*, 75–8; Blackmore, *Hunting Weapons*, 205; Jens Sensfelder, 'Technische Geschichte Der Armbrust am Beispiel Der Berliner Sammlung', in *Die Armbrust Schrecken und Schönheit*, ed. Sven Lüken and Jens Sensfelder (Munich, 2019), 83–8.
23. Blackmore, *Hunting Weapons*, 164, 205–08; Paterson, *A Guide to the Crossbow*, 92–5; Alm, *European Crossbows*, 93–4.
24. Breiding, *A Deadly Art*, 66–78; Payne-Gallwey, *The Crossbow*, 177.
25. Breiding, *A Deadly Art*, 67–8; Alm, *European Crossbows*, 95–6.
26. Breiding, *A Deadly Art*, 82–3, 90–1; Paterson, *A Guide to the Crossbow*, 105–07.
27. Richter, *Die Hornbogenarmbrust*, 90–2; Liebel, *Springalds and Great Crossbows*, 22–7.
28. Liebel, *Springalds and Great Crossbows*, 23–9, 42–3.

29. Ibid., 30–2, 35–6, 42–8.
30. Ibid., 36–43, 50–2, 55–6.
31. Alm, *European Crossbows*, 90–2.
32. Ibid., 90–2; Gorman, *Technological Development of the Bow and the Crossbow*, 176.
33. Breiding, *A Deadly Art*, 80; Blackmore, *Hunting Weapons*, 198–9; Paterson, *A Guide to the Crossbow*, 99–102.
34. Breiding, *A Deadly Art*, 80–1, 84–6.
35. Blackmore, *Hunting Weapons*, 199, 204.
36. Paterson, *A Guide to the Crossbow*, 101.

Chapter 3

1. Robert Ignatius Burns, 'The Medieval Crossbow as Surgical Instrument: An Illustrated Case History', *Bulletin of the New York Academy of Medicine* 48;8 (1972), 985–6.
2. Ibid., 987–8; Henri de Mondeville, *Chirurgie de Maitre Henri de Mondeville, Chirurgien de Philippe le Bel, Roi de France, Composee de 1306 a 1320*, trans. E. Nicais (Paris, 1893), 235–7.
3. Jim Bradbury, *The Medieval Archer* (Woodbridge, 1985), 76–8; Gorman, *Technological Development of the Bow and the Crossbow*, 137–42.
4. Clifford Rogers, 'The development of the longbow in late medieval England and "technological determinism"', *Journal of Medieval History* 37;3 (2011), 327, 336–40; Enrico Ascani and Francesco Gorgo, 'A Technical Study of the Fresco in the "St Sebastian's Chapel" Celle Macra, Italy', *Journal of the Society of Archer-Antiquaries* 42 (1999), 86; Enrico Lazzari, 'The 15th Century Crossbow in the Burgundy and Italian Iconography: Analysis and Assumptions', *Jahrblatt der Interessengemeinschaft Historische Armbrust* 7 (2015), 95–105.
5. Edward Williams Byron Nicholson, *Keltic Researches: Studies in the History and Distribution of the Ancient Goidelic Language and Peoples* (London, 1904), 74–5.
6. Blackmore, *Hunting Weapons*, 178.
7. Paterson, *A Guide to the Crossbow*, 16–17.
8. E. Esperandieu, *Recueil general des bas-reliefs de La Gaule romaine*, 12 vols (Paris, 1907–28), 2: ID nos. 1679, 1683.
9. Ibid., 2;443–4; Joseph Needham, Robert D.S. Yates, Krysztof Gawlikowski, Edward McEwen and Wang Ling, *Science and Civilisation in China, Volume 5. Chemistry and Chemical Technology, Part 6. Military Technology: Missiles and Sieges* (Cambridge, 1998), 123–5, 127–32, 141–6; Pétrin, 'Philological notes on the crossbow and related missile weapons', 277–8.
10. Esperandieu, *Recueil general des bas-reliefs de La Gaule romaine*, 2;442; Blackmore, *Hunting Weapons*, 174; Pétrin, 'Philological notes on the crossbow and related missile weapons', 277–8.
11. Soria, Cathedral Library Burgo de Osma, Cod. 1, fol. 85v.
12. Hunting Scene, Anonymous, Metropolitan Museum of Art, New York City, Accession No. 57.97.2.
13. Daniel Weiss, 'Portraying the Past, Illuminating the Present: The Art of the Morgan Library Picture Bible', in *The Book of Kings: Art, War, and the Morgan Library's Medieval Picture Bible*, ed. William Noel and Daniel Weiss (Baltimore, 2002), 14.
14. New York City, Pierpont Morgan Library, Ms M. 638; Paris, Bibliothèque nationale de France, MS nouv. acq. lat. 2294; Los Angeles, J. Paul Getty Museum, MS 16.
15. Ian Pierce, 'The Knight, his Arms and Armour c.1150-1250', in *Anglo Norman Studies XV The Battle Conference*, ed. Marjorie Chibnall (Woodbridge, 1992), 253, 270; R. Ewart Oakeshott, *The Archaeology of Weapons Arms and Armour from Prehistory to the Age of Chivalry* (London, 1960; repr. 1996), 262–3. Stephen N. Fliegel, 'The Art of War: Thirteenth Century Arms and Armour', in *The Book of Kings: Art, War, and the Morgan Library's Medieval Picture Bible,* ed. William Noel and Daniel Weiss (Baltimore, 2002), 83–4; Matthew Strickland and Robert Hardy, *The Great Warbow: From Hastings to the Mary Rose* (Stroud, 2005; repr. 2011), 46.
16. New York City, Pierpont Morgan Library, Ms M. 638, fol. 10v.
17. New York City, Pierpont Morgan Library, Ms M. 638, fol. 42r.
18. Oxford, Bodleian Ms. 264, fol. 81v.
19. Dublin, Chester Beatty Library, W 082, fol. 317r.
20. Paris, Bibliothèque nationale de France, Fr 2643, f.165v.

21. Battle of Higueruela by Fabrizio Castello, Orazio Cambiasi and Lazzaro Tavarone in the Gallery of Battles at the Royal Monastery of San Lorenzo de El Escorial.
22. Master ES, Martyrdom of St Sebastian, c.1450/1460, National Museum of Art, Washington, DC.
23. Attributed to Cherubino Alberti, An Archer Shooting a Crossbow, 1579, National Gallery, Washington, DC; After Jacques de Gheyn II, The Archer and the Milkmaid, c.1610, Metropolitan Museum of Art, New York City.
24. Attributed to Gerhart Altzenbach, A Skeleton with Bow and Arrow, before 1673, Wellcome Collection, London.
25. Meister der Heiligen Sippe d. J, Martyrdom of St Sebastian, c.1480, Bildarchiv & Wallraf-Richartz Museum, Cologne.
26. Piero de Pollaiuolo, Martyrdom of St Sebastian, c.1475, National Gallery, London.
27. Lazzari, 'The 15th Century Crossbow in the Burgundy and Italian Iconography: Analysis and Assumptions', 97.
28. Hans Holbein the Elder, Martyrdom of St Sebastian, 1516, Alte Pinakothek, Munich.
29. Pietro Perugino, Martyrdom of St Sebastian, 1505, Church of St Sebastian, Panicale.
30. Lazzari, 'The 15th Century Crossbow in the Burgundy and Italian Iconography: Analysis and Assumptions', 99.
31. Olaus Magnus, *Historia de Gentibus Septentrionalibus* (Rome, 1555), 231.
32. Hans Sachs and Jost Amman, *Das Ständebuch* (Frankfurt am Main, 1568), Quire L6.
33. Serdon, *Armes Du Diable*, 185–6; Holger Richter, *Die Kunst der Armbrustmacher in Dresden* (Beucha, 2008), 87.
34. Breiding, *A Deadly Art*, 21–2.
35. Ibid., 23–6.
36. Ibid., 26–9.
37. Ibid., 35–6.
38. Stuart Gorman, *Technological Development of the Bow and the Crossbow*, 290–1.
39. Crossbow, 1530/60, reference number 1982.3073, Art Institute Chicago.

Chapter 4

1. Geoffrey Gaimar, *LEstorie de Engles*, ed. and trans. Thomas Duffus Hardy and Charles Trice Martin, 2 vols (London, 1889), 2: lines 4408–4428.
2. Ibid., 2: lines 4447–4470.
3. Henry of Huntingdon, *The Chronicle of Henry of Huntingdon*, 196.
4. *Cnutonis Regis Gesta; Sive, Encomium Emmae Reginae, Auctore Monacho Sancti Bertini*, ed. Georg Heinrich Pertz (Hanover, 1865), 22–3.
5. Needham et al, *Science and Civilisation in China*, 123–5, 127–32, 141–6; Xiuzhen Janice Li, Andrew Bevan, Marcos Martinón-Torres, Thilo Rehren, Wei Cao, Yin Xia and Kun Zhao, 'Crossbows and Imperial Craft Organisation: the Bronze Triggers of China's Terracotta Army', *Antiquity* 88, no. 339 (2014), 126–38; news article on complete crossbow: https://www.archaeology.org/news/3105-150320-china-qin-crossbow
6. Needham et al, *Science and Civilisation in China*, 100–200; Mike Loades, *The Crossbow* (Oxford, 2018), 7–13; Paterson, *A Guide to the Crossbow*, 53–4; Gorman, *Technological Development of the Bow and the Crossbow*, 184–5.
7. Sun Tzu, *The Art of War*, ed. and trans. Ralph D. Sawyer (Boulder, 1994), 187.
8. Needham et al, *Science and Civilisation in China*, 135–9.
9. Ibid., 139–40, 144–6; Li et al, 'Crossbows and Imperial Craft Organisation', 129–30; David Curtis Wright, 'Nomadic Power, Sedentary Security, and the Crossbow', in John Ferris (ed.), *Calgary Papers in Military and Strategic Studies* 2 (Calgary, 2008), 79–86, 73–90.
10. Needham et al, *Science and Civilisation in China*, 170–6.
11. Loades, *The Crossbow*, 14–15; Paterson, *A Guide to the Crossbow*, 13–14; Needham et al, *Science and Civilisation in China*, 170–1.
12. Robert L. O'Connell, 'The Life and Hard Times of the Crossbow', *MHQ The Quarterly Journal of Military History* 1;2 (1989), 48.

13. Chevedden, 'Artillery in Late Antiquity', 138–52; Pétrin, 'Philological notes on the crossbow and related missile weapons', 269–70; Needham et al, *Science and Civilisation in China*, 171–2.
14. Chevedden, 'Artillery in Late Antiquity ', 138–52.
15. Paterson, *A Guide to the Crossbow*, 16–19.
16. Loades, *The Crossbow*, 14–16.
17. Bernard S. Bachrach, *Charlemagne's Early Campaigns (768-777): A Diplomatic and Military Analysis* (Leiden, 2013), 120, 233, 324, 352, 486.
18. Peter Purton, *A History of the Early Medieval Siege c.450-1200* (Woodbridge, 2009), 83–5.
19. Anthony Adams and A.G. Rigg, 'A Verse Translation of Abbo of St Germain's *Bella Parisciacae Urbis*', *Journal of Medieval Latin* 14 (2004), 30.
20. Ibid., 24–6, 28.
21. David S. Bachrach, *Warfare in Tenth-Century Germany* (Woodbridge, 1998), 155–6.; Richer of Rheims, *Histoire de son temps*, ed. and trans. J. Gaudet, 2 vols (Paris, 1845), 1:64–5.
22. Purton, *A History of the Early Medieval Siege*, 147–8; Richer of Rheims, *Histoire de son temps*, 2:124–35.
23. Pétrin, 'Philological notes on the crossbow and related missile weapons', 284–6.

Chapter 5

1. Orderic Vitalis, *The Ecclesiastical History of England and Normandy*, trans. Thomas Forrester, 4 vols. (London, 1853), 3:465–7.
2. Ibid., 3: 467.
3. Ibid., 4:19.
4. Flynn, *Crusader Artillery, 1097-1148* , iv, 29–30.
5. Ibid., 67–8.
6. Thomas Asbridge, *The Crusades: The War for the Holy Land* (London, 2010), 52–6.
7. Flynn, *Crusader Artillery, 1097-1148*, 111–13.
8. Ibid., 130, 163; *Chronicles of the First Crusade*, ed. and trans. Chris Tyerman (London, 2004; repr. 2012). 112
9. Flynn, *Crusader Artillery, 1097-1148*, 125–7; *Chronicles of the First Crusade*, 110; Peter Purton, *The Medieval Military Engineer: From the Roman Empire to the Sixteenth Century* (Woodbridge, 2018), 131–3, 142–8.
10. Asbridge, *The Crusades*, 62–88.
11. Flynn, *Crusader Artillery, 1097-1148*, 172–3.
12. Ibid., 86–7.
13. Asbridge, *The Crusades*, 89–102.
14. Steve Tibble, *The Crusader Armies: 1099-1187* (New Haven, 2020), 159–64.
15. Ibid., 3–8, 37–50.
16. John D. Hosler, *The Siege of Acre* (New Haven, 2020), 42.
17. Ibid., 177–80.
18. Asbridge, *The Crusades*, 412–27.
19. Ibid., 430–52; Hosler, *The Siege of Acre*, 131–43; Bahā' al-Dīn Ibn Shaddād, *The Rare and Excellent History of Saladin*, trans. D.S. Richards (Aldershot, 2002), 161, 164–5.
20. *Chronicles of the Crusades*, trans. J.A. Giles and Thomas Jones (London, 1848), 117–18.
21. Hosler, *The Siege of Acre*, 56, 79–82; Bahā' al-Dīn Ibn Shaddād, *The Rare and Excellent History of Saladin*, 127–8.
22. Hosler, *The Siege of Acre*, 110, 120; Ambroise, *The History of the Holy War*, trans. Marianne Ailes (2003, repr. Woodbridge, 2011), 99.
23. Asbridge, *The Crusades*, 398–410; Hosler, *The Siege of Acre*, 27–38; Bahā' al-Dīn Ibn Shaddād, *The Rare and Excellent History of Saladin*, 98–9, 101–05.
24. Flynn, *Crusader Artillery, 1097-1148*, 86–7; Ambroise, *The History of the Holy War*, 82.
25. Ambroise, *The History of the Holy War*, 84, 100–01.
26. Ibid., 108.
27. Tibble, *The Crusader Armies*, 128–9, 138–49.
28. Ibid., 140–1, 143–9.
29. Asbridge, *The Crusades*, 462–4; Strickland and Hardy, *The Great Warbow*, 106.

30. Asbridge, *The Crusades*, 465–6; Ambroise, *The History of the Holy War*, 115.
31. Tibble, *The Crusader Armies*, 143–9; Bahā' al-Dīn Ibn Shaddād, *The Rare and Excellent History of Saladin*, 174–6.
32. Asbridge, *The Crusades*, 466–74; Ambroise, *The History of the Holy War*, 115–21.
33. Asbridge, *The Crusades*, 510; Ambroise, *The History of the Holy War*, 179.
34. Strickland and Hardy, *The Great Warbow*, 107; Ralph of Coggeshall, *Chronicon Anglicanum*, 50; Tibble, *The Crusader Armies*, 150–4; Ambroise, *The History of the Holy War*, 175–85; Bahā' al-Dīn Ibn Shaddād, *The Rare and Excellent History of Saladin*, 217–23.
35. Asbridge, *The Crusades*, 512.
36. Flynn, *Crusader Artillery, 1097-1148*, 81.
37. Strickland and Hardy, *The Great Warbow*, 114.
38. 'Second Lateran Council – 1139 AD', Papal Encyclicals Online, accessed 29 May 2021, https://web.archive.org/web/20210529124452/https://www.papalencyclicals.net/councils/ecum10.htm
39. Ibid.
40. Strickland and Hardy, *The Great Warbow*, 116–18.
41. O'Connell, 'The Life and Hard Times of the Crossbow', 48; Jonathan Riley-Smith, *The First Crusade and the Idea of Crusading* (New York, 1986; repr. 2003), 2–5; Philippe Contamine, *War in the Middle Ages*, trans. Michael Jones (Oxford, 1984), 274–5.
42. Grummitt, *The Calais Garrison*, 45–52, 120–1; N.J.C. Smith, *Servicium Debitum and Scutage in Twelfth-Century England with Comparisons to the Regno of Southern Italy*, unpublished thesis (2010), 149–52.
43. John France, *Western Warfare in the Age of the Crusades, 1000-1300* (Ithaca, 1999), 26–7; Hall, *Weapons and Warfare in Renaissance Europe*, 16–17.
44. Thomas Asbridge, *The Greatest Knight* (London, 2015), 351–63.
45. Ibid., 351–63; *The History of William Marshal*, trans. Nigel Bryant (Woodbridge, 2016), 195–204.
46. *The History of William Marshal*, 201, 205–11; Roger of Wendover, *Flowers of History*, translated by J.A. Giles, 2 vols (London, 1849), 2:389–98.
47. Matthew Paris, *Chronica Majora*, ed. Henry Richards Luard, 7 vols (London, 1876), 3:19; Strickland and Hardy, *The Great Warbow*, 134.
48. Strickland and Hardy, *The Great Warbow*, 441.
49. Steven Runciman, *The Sicilian Vespers* (New York, 1958; repr. 1960), 81–92.
50. Ibid., 93–104
51. Ibid., 104–10; France, *Western Warfare in the Age of the Crusades*, 178–80; Strickland and Hardy, *The Great Warbow*, 114; Charles Oman, *A History of the Art of War: the Middle Ages from the Fourth to the Fourteenth Century* (London, 1898), 480–3; *Annales Ianuenses*, ed. Georg Heinrich Pertz (Hanover, 1862), 255–8; Michael Mallett, *Mercenaries and Their Masters* (London, 1975; repr. 2009), 15.
52. Runciman, *The Sicilian Vespers*, 109–12; Mallett, *Mercenaries and Their Masters*, 8; Oman, *A History of the Art of War*, 485–8; *Die Chronik des Saba Malaspina*, trans. Walter Koller and August Nitschke (Hanover, 1999), 164–74; Andrew of Hungary, *Descriptio Victoriae a Karolo Provinciae comite reportate*, ed. G. Waitz (Hanover, 1882), 575–7.
53. France, *Western Warfare in the Age of the Crusades*, 180.

Chapter 6

1. Peter Purton, *A History of the Late Medieval Siege 1200-1500* (Woodbridge, 2010), 259–61; Jim Bradbury, *The Medieval Siege* (Woodbridge, 1992), 172–5.
2. Kelly DeVries, *Joan of Arc: A Military Leader* (Stroud, 1999), 50–9; Craig Taylor (ed. and trans.), *Joan of Arc: La Pucelle* (Manchester, 2006), 84, 158, 280–1, 287, 296, 315–17, 357.
3. Taylor, *Joan of Arc*, 256–7.
4. Ibid., 230.
5. Ibid., 125, 145, 199, 253; Purton, *A History of the Late Medieval Siege 1200-1500*, 261; DeVries, *Joan of Arc*, 129–36, 140–7.
6. DeVries, *Joan of Arc*, 151, 163–9; Bradbury, *The Medieval Siege*, 174.
7. J.F. Verbruggen, *The Art of Warfare in Western Europe* (1954, repr. Woodbridge, 1997), 111–12.
8. Kelly DeVries, *Infantry Warfare in the Early Fourteenth Century* (Woodbridge, 1996), 9–10.

9. Ibid., 12–13; Verbruggen, *The Art of Warfare in Western Europe*, 190.
10. Verbruggen, *The Art of Warfare in Western Europe*, 192–3.
11. Strickland and Hardy, *The Great Warbow*, 135; DeVries, *Infantry Warfare in the Early Fourteenth Century*, 17–22; Verbruggen, *The Art of Warfare in Western Europe*, 192–4.
12. Strickland and Hardy, *The Great Warbow*, 135; DeVries, *Infantry Warfare in the Early Fourteenth Century*, 32–48; Verbruggen, *The Art of Warfare in Western Europe*, 198–203.
13. J.F. Verbruggen, 'Flemish Urban Militias against the French Cavalry Armies in the Fourteenth and Fifteenth Centuries', *Journal of Medieval Military History* 1 (2002), 159–63, 167–8.
14. David Green, *The Hundred Years War: A People's History* (New Haven, 2014), 26–30; Andrew Ayton, 'The Battle of Crécy: Context and Significance', in *The Battle of Crécy, 1346*, ed. Andrew Ayton and Sir Philip Preston (Woodbridge, 2005), 18–20.
15. J.E. Morris, 'The Archers at Crécy', *The English Historical Review* 12;47 (1897), 427–36; Oman, *A History of the Art of War*, 102–05.
16. Andrew Ayton, 'Crécy and the Chroniclers', in *The Battle of Crécy, 1346*, ed. Andrew Ayton and Sir Philip Preston (Woodbridge, 2005), 287–350; DeVries, *Infantry Warfare in the Early Fourteenth Century*, 158–63.
17. Andrew Ayton, 'The Crécy Campaign', in *The Battle of Crécy, 1346*, ed. Andrew Ayton and Sir Philip Preston (Woodbridge, 2005), 35–41; DeVries, *Infantry Warfare in the Early Fourteenth Century*, 155–8.
18. DeVries, *Infantry Warfare in the Early Fourteenth Century*, 165–7; Jean le Bel, *The True Chronicles of Jean le Bel*, trans. Nigel Bryant (Woodbridge, 2011), 179–80; Jean Froissart, *Chronicles*, trans. Geoffrey Brereton (London, 1968; repr. 1978), 88–9; Geoffrey le Baker, *The Chronicle of Geoffrey le Baker*, trans. David Preest and ed. Richard Barber (Woodbridge, 2012), 72; Giovanni Villani, 'New Chronicle', in *The Battle of Crécy: A Casebook*, ed. Michael Livingston and Kelly DeVries (Liverpool, 2015), 117.
19. DeVries, *Infantry Warfare in the Early Fourteenth Century*, 167–74.
20. Jean le Bel, *The True Chronicles of Jean le Bel*, 179–80; Froissart, *Chronicles*, 88–9; Geoffrey le Baker, *The Chronicle of Geoffrey le Baker*, 73.
21. Oman, *A History of the Art of War*, 102–05; J.E. Morris, *The Welsh Wars of Edward I* (Oxford, 1901; repr. 1996), 26–34, 87–94, 99–104.
22. 'Grandes Chroniques', in *The Battle of Crécy: A Casebook*, ed. Michael Livingston and Kelly DeVries, (Liverpool, 2015), 130–2; Jean de Venette, 'Chronicles', in *The Battle of Crécy: A Casebook*, ed. Michael Livingston and Kelly DeVries (Liverpool, 2015), 215; DeVries, *Infantry Warfare in the Early Fourteenth Century*, 167–8.
23. Payne-Gallwey, *The Crossbow*, 4–7; Paterson, *A Guide to the Crossbow*, 113–22.
24. Kelly DeVries, 'The Tactics of Crécy', in *The Battle of Crécy: A Casebook*, ed. Michael Livingston and Kelly DeVries (Liverpool, 2015), 455.
25. Villani, 'New Chronicle', 117.
26. DeVries, 'The Tactics of Crécy', 452–4; Michael Prestwich, 'The Battle of Crécy', in *The Battle of Crécy, 1346*, ed. Andrew Ayton and Sir Philip Preston (Woodbridge, 2005), 148; Bradbury, *The Medieval Archer*, 105–09.
27. Russell Mitchell, 'The Longbow-Crossbow Shootout at Crécy (1346): Has the "Rate of Fire Commonplace" Been Overrated?', in *The Hundred Years War (Part II) Different Vistas*, ed. L.J. Andrew Villalon and Donald J. Kagay (Leiden, 2008), 235–49; Hall, *Weapons and Warfare in Renaissance Europe*, 19–20.
28. Grummitt, *The Calais Garrison*, 105; Kelly DeVries and Niccolò Capponi, 'The Genoese Crossbowmen at Crécy', in *The Battle of Crécy: A Casebook*, ed. Michael Livingston and Kelly DeVries (Liverpool, 2015), 441–6.
29. João Gouveia Monteiro, 'The Battle of Aljubarrota (1385): A reassessment', *Journal of Medieval Military History* 7 (2009), 76–9; Robert Bartlett, *Blood Royal: Dynastic Politics in Medieval Europe* (Cambridge, 2020), 172–3; Peter Ainsworth, and Godfried Croenen, ed., 'Translation of Book II'. *The Online Froissart*, Accessed 25 July 2021. https://www.dhi.ac.uk/onlinefroissart/browsey.jsp?pb0=-BookII-Translation_67v&img0=&div0=ms.f.transl.BookII-Translation&panes=1&GlobalMode=facsimile&img0=&disp0=pb&GlobalWord=0&GlobalShf=&pb0=BookII-Translation_68r
30. Monteiro, 'The Battle of Aljubarrota (1385): A reassessment', 79–92.
31. Ibid., 92–103.

32. Ibid., 89, 98; Anne Curry, *Great Battles: Agincourt* (Oxford, 2015), 12–14; Mallett, *Mercenaries and Their Masters*, 54–5; Strickland and Hardy, *The Great Warbow*, 251–4.
33. Mike Carr, *Merchant Crusaders in the Aegean 1291-1352* (Woodbridge, 2015), 81–3.
34. Frederic C. Lane, 'The Crossbow in the Nautical Revolution of the Middle Ages', *Explorations in Economic History* 7;1-2 (1969), 161–9.
35. Dirk Meier, *Seafarers, Merchants and Pirates in the Middle Ages*, trans. Angus McGeoch (Woodbridge, 2006), 33–40; Frances Gies and Joseph Gies, *Cathedral, Forge, and Waterwheel: Technology and Innovation in the Middle Ages* (New York, 1994), 154–6, 222.
36. Geoffrey le Baker, *The Chronicle of Geoffrey le Baker*, 95.
37. Graham Cushway, *Edward III and the War at Sea: The English Navy, 1327-1377* (Woodbridge, 2011), 136–42.
38. Andrew Ayton and Craig Lambert, 'Navies and Maritime Warfare', in *The Hundred Years War Revisited*, ed. Anne Curry (London, 2019), 174–7, 201–03.
39. Eric Christiansen, *The Northern Crusades* (New York, 1980; repr. 1997), 7–20, 67, 91.
40. Ekdahl, 'Horses and Crossbows', 137.
41. William Urban, *Last Years of the Teutonic Knights* (Barnsley, 2018), 32, 42, 119–23; Ekdahl, 'Horses and Crossbows', 149–50.
42. Ekdahl, 'Horses and Crossbows', 136.
43. Ibid., 148; Nowakowski, *Arms and Armour in the Medieval Teutonic Order's State in Prussia*, 96–102.
44. Christiansen, *The Northern Crusades*, 67, 101.
45. Ibid., 91, 106–07, 112.
46. Ibid., 167–76; Nowakowski, *Arms and Armour in the Medieval Teutonic Order's State in Prussia*, 101.
47. Christiansen, *The Northern Crusades*, 140; Rackevicius, 'The Crossbow – The Weapon of the Invaders and The Defenders of Vilnius Castle (The Late 14th – Early 15th Centuries)', 59–63; Urban, *Last Years of the Teutonic Knights*, 107.
48. 'The Very Pretty Chronicle of John Žižka The Servant of King Wenceslas', in Frederick G. Heymann, *John Žižka and the Hussite Revolution* (Princeton, 1955), 8.
49. Heymann, *John Žižka and the Hussite Revolution*, 61–74, 148–63; Urban, *The Last Years of the Teutonic Knights*, 172–3, 206–09.
50. Heymann, *John Žižka and the Hussite Revolution*, 16–35.
51. Hall, *Weapons and Warfare in Renaissance Europe*, 108; Heymann, *John Žižka and the Hussite Revolution*, 97–100, 366.
52. Ekdahl, 'Horses and Crossbows', 151; Hall, *Weapons and Warfare in Renaissance Europe*, 112; DeVries, *Infantry Warfare in the Early Fourteenth Century*, 97–100.
53. DeVries, *Infantry Warfare in the Early Fourteenth Century*, 124–5.
54. Ibid., 408–09.
55. Ibid., 292–304; Lawrence of Březová, 'Vavřince Z Březové Kronika Husitská', in *Fontes Rerum Bohemicarum: Prameny Dějin Českých*, ed. Josephus Emler, 8 vols (Prague, 1871–1932), 5:532–4.
56. Villani, 'New Chronicle', 116–17; Ambroise, *The History of the Holy War*, 175–85.
57. Heymann, *John Žižka and the Hussite Revolution*, 90.
58. Hall, *Weapons and Warfare in Renaissance Europe*, 110; Urban, *The Last Years of the Teutonic Knights*, 209–14; Heymann, *John Žižka and the Hussite Revolution*, 98–100, 174–5, 284–5, 366–7.
59. Hall, *Weapons and Warfare in Renaissance Europe*, 169.
60. Strickland and Hardy, *The Great Warbow*, 255.
61. David Potter, *Renaissance France at War* (Woodbridge, 2008), 97; Contamine, *War in the Middle Ages*, 217.
62. Strickland and Hardy, *The Great Warbow*, 254–7.
63. Ibid., 354; Mallett, *Mercenaries and Their Masters*, 108; Potter, *Renaissance France at War*, 70–2, 78–80.
64. Strickland and Hardy, *The Great Warbow*, 354.
65. Potter, *Renaissance France at War*, 100.
66. Strickland and Hardy, *The Great Warbow*, 354–5.
67. Ibid., 355.
68. Hall, *Weapons and Warfare in Renaissance Europe*, 114–23.
69. Potter, *Renaissance France at War*, 80.
70. Ibid., 102–05, 324–7.

71. Ibid., 80.
72. Ibid., 79–80, 227–8, 233–5.
73. Mallett, *Mercenaries and Their Masters*, 238; Hall, *Weapons and Warfare in Renaissance Europe*, 171–4.
74. Liebel, *Springalds and Great Crossbows*, 56; Hall, *Weapons and Warfare in Renaissance Europe*, 184.
75. Alm, *European Crossbows*, 70–8.
76. Ibid., 70–1.
77. Ibid., 72.
78. Ibid.
79. Ernle Bradford, *The Great Siege. Malta 1565* (London, 1961; repr. 2014), 14–15; Roger Crowley, *Empires of the Sea: The Final Battle for the Mediterranean 1521-1580* (London, 2008), 50, 106–10.
80. Bradford, *The Great Siege*, 14–16; Theresa M. Vann, 'The Militia of Malta', *The Journal of Medieval Military History* 2 (2004), 146–8.
81. Bradford, *The Great Siege*, 32–7.
82. Ibid., 47–9.
83. Ibid., 40–1.
84. Ibid., 40, 53–60, 63–4; Francesco Balbi di Correggio, *The Siege of Malta 1565*, trans. Ernle Bradford (London, 1965; repr. 2005), 45–51.
85. Bradford, *The Great Siege*, 66–74, 86; Balbi di Correggio, *The Siege of Malta 1565*, 51–4.
86. Bradford, *The Great Siege*, 108–13, 120–3; Balbi di Correggio, *The Siege of Malta 1565*, 68–89.
87. Bradford, *The Great Siege*, 138–40; Balbi di Correggio, *The Siege of Malta 1565*, 89–104.
88. Bradford, *The Great Siege*, 126, 132–3, 138–40.
89. Ibid., 144–6; Balbi di Correggio, *The Siege of Malta 1565*, 102–03.
90. Bradford, *The Great Siege*, 154–9; Balbi di Correggio, *The Siege of Malta 1565*, 111–18.
91. Bradford, *The Great Siege*, 170–5.
92. Crowley, *Empires of the Sea*, 183; Balbi di Correggio, *The Siege of Malta 1565*, 161–2; Francisco Balbi di Correggio, *La Verdadera Relacion de Todo Lo Que El Anno De M.D.LXV.* (Barcelona, 1568), 100–01; Stephen C. Spiteri, *Armoury of the Knights* (Valletta, 2002), 57.
93. Bradford, *The Great Siege*, 211, 216–21.
94. Spiteri, *Armoury of the Knights*, 54.
95. Ibid., 54, 57, 123, 257; Guy Francis Laking, *A catalogue of the armour and arms in the Armoury of the Knights of St John of Jerusalem, now in the Palace, Valletta, Malta* (London, 1903), 147.
96. Balbi di Correggio, *La Verdadera Relacion de Todo Lo Que El Anno De M.D.LXV*, 109.
97. Crowley, *Empires of the Sea*, 281.
98. Arthur G. Credland, 'The Crossbow and the Bow in Modern Warfare', *Arms and Armour* 7;1 (2010), 55–74.

Chapter 7

1. Jenny Benham, *Peacemaking in the Middle Ages* (Manchester, 2010), 158.
2. Jorgan Ebert, 'Gamle Danske Sange: Til Refsnæs Kong Valdemar red på jagt.' ugle.dk. https://web.archive.org/web/20210725212851/https://ugle.dk/til_refsnaes_kong_valdemar.html (Accessed: 25/07/21); Otto Bloom, 'Blidemestere, Balistarii og Værkmestere i Kjøbenhavn ca. 1375—1550', *Historisk Tidsskrift* 5;5 (1885), 74.
3. 'Chronica Jutensis', in *Scriptores Minores Historiæ Danicæ*, ed. C.L. Gertz, 2 vols (Copenhagen, 1917–20), 1:443.
4. 'Vetus Chronica Sialandie', in *Scriptores Minores Historiæ Danicæ*, ed. C.L. Gertz, 2 vols (Copenhagen, 1917–20), 2:64.
5. Ebert, 'Gamle Danske Sange: Til Refsnæs Kong Valdemar red på jagt.'.
6. C. Warren Hollister, 'The Strange Death of William Rufus', *Speculum* 48; 4 (1973), 640–8.
7. Blackmore, *Hunting Weapons*, 175.
8. Niketas Choniates, *O City of Byzantium, Annals of Niketas Choniates*, trans. Harry J. Magoulias (Detroit, 1984), 26.
9. An Smets and Baudouin Van Den Abeele, 'Medieval Hunting', in *A Cultural History of Animals in the Medieval Age*, ed. Brigitte Resl (Oxford, 2007), 59–64.

10. David Scott-Mcnab, 'Richard Almond: *Medieval Hunting* [Book Review]', *Envoi A Review Journal of Medieval Literature* 11;1 (2002), 23; Smets and Van Den Abeele, 'Medieval Hunting', 65–71.
11. Blackmore, *Hunting Weapons*, 175–85; Gorman, *Technological Development of the Bow and the Crossbow*, 147–50, 176–7.
12. Payne-Gallwey, *The Crossbow*, 129–68; Blackmore, *Hunting Weapons*, 185–6; Gorman, *Technological Development of the Bow and the Crossbow*, 169–79.
13. Blackmore, *Hunting Weapons*, 197–205; Gorman, *Technological Development of the Bow and the Crossbow*, 169–79.
14. Froissart, *Chronicles*, 268–74; Rebekah Pratt-Sturges, *Illuminating the Medieval Hunt: Power and Performance in Gaston Fébus' Le livre de Chasse*, unpublished thesis (2017), 1–3.
15. Pratt-Sturges, *Illuminating the Medieval Hunt*, 5–6.
16. Ibid., 81–2; Gaston Phoebus, *Le Chasse de Gaston Phoebus*, ed. and trans. Jean de Foix and Joseph Lavallée (Paris, 1854), 30, 215, 260.
17. Paris, Bibliothèque nationale de France, MS fr. 616, fol. 111v.
18. Paris, Bibliothèque nationale de France, MS fr. 616, fol. 115r.
19. Gaston Phoebus, *The Hunting Book*, ed. Gabriel Bise and trans. J. Peter Tallon (London, 1978), 95.
20. Paris, Bibliothèque nationale de France, MS fr. 616, fol. 113v.
21. Paris, Bibliothèque nationale de France, MS fr. 616, fol. 114r.
22. Gaston Phoebus, *The Hunting Book*, 96.
23. Paris, Bibliothèque nationale de France, MS fr. 616, fol. 117r.
24. Gaston Phoebus, *The Hunting Book*, 97.
25. Ibid., 76.
26. Paris, Bibliothèque nationale de France, MS fr. 616, fol. 93r.
27. Gaston Phoebus, *Le Chasse de Gaston Phoebus*, 215; Pratt-Sturges, *Illuminating the Medieval Hunt*, 119–20.
28. Edward Norwich, *The Master of Game*, ed. and trans. William. A. Baillie-Grohman (New York, 1909), xi–xviii, 27, 30, 223, 237.
29. Arthur G. Credland, 'Aquatic Hunting with the Crossbow', *Jahrblatt der Interessengemeinschaft Historische Armbrust* 5 (2013), 33.
30. Ibid., 33–42.
31. Ibid., 43–5.
32. Ibid; Alm, *European Crossbows*, 9–10, 89–90.
33. Alm, *European Crossbows*, 9–10, 89–90.
34. Ibid., 80–1; Olaus Magnus, *Historia de Gentibus Septentrionalibus*, 501–02.
35. Alm, *European Crossbows*, 31; Payne-Gallwey, *The Crossbow*, 224–5; Laura Crombie, *Archery and Crossbow Guilds in Medieval Flanders, 1300-1500* (Woodbridge, 2018), 89–93, 160–1.
36. Alm, *European Crossbows*, 31–2; Crombie, *Archery and Crossbow Guilds in Medieval Flanders*, 152–8; Potter, *Renaissance France at War*, 100.
37. Alm, *European Crossbows*, 31; Crombie, *Archery and Crossbow Guilds in Medieval Flanders*, 89–93.
38. Payne-Gallwey, *The Crossbow*, 231–6.
39. Gian Paolo G. Scharf, 'Il <ludus balistarum> nel Medio Evo, con particular riguardo all'area ligure-piemontese', in *Giochi e Giocattoli Nel Medieoevo Piemontese e Ligure*, ed. Rinaldo Comba and Riccardo Rao (Rocca de'Baldi, 2005), 217–19; For the Holy Roman Empire see J.-D. Delle Luche, *Des amitiés ciblées: Concours de tir et diplomatie urbaine dans le Saint-Empire, XVe-XVIe siècle* (Turnhout, 2021). For Switzerland see Walter Schaufelberger, *Der Wettkampf in der alten Eidgenossenschaft* (Bern, 1972).
40. Crombie, *Archery and Crossbow Guilds in Medieval Flanders*, 22–37.
41. Ibid., 160–89; Laura Crombie, 'Representatives of Civic Pride and Cultural Identities; the Ghent Crossbow Competition of 1498', *Arms and Armour* 8;2 (2011), 159–61.
42. Sergio Boffa, *Warfare in Medieval Brabant* (Woodbridge, 2004), 144–51.
43. Crombie, *Archery and Crossbow Guilds in Medieval Flanders*, 153–4.

Chapter 8

1. Friedrich Schiller, *Schiller Volume Three: Joan of Arc, William Tell*, trans. Robert David MacDonald (London, 2005), 139–245.
2. Aegidius Tschudi, *Chronicon Helveticum*, ed. and trans. Bernhard Stettler, 22 vols (Basel, 1970–2001), 1:346–51; 1:444–58.
3. Hans Schriber, *Das Weisse Buch von Sarnen*, ed. and trans. Bruno Meyer (Sarnen, 1984), 43–5.
4. Sabine Baring-Gould, *Curious Myths of the Middle Ages* (London, 1877), 113–33; Jean-François Bergier, *Wilhelm Tell, Realität und Mythos* (Munich, 1988), 371–2.
5. Ludwig Tobler, *Schweizerische Volkslieder* (Frauenfeld, 1882), 3–5; E.L. Rochholz, *Tell und Gessler in Sage und Geschichte* (Heilbronn, 1877), 180–7.
6. Hall, *Weapons and Warfare in Renaissance Europe*, 37, 122; Richard Vaughan, *Charles the Bold* (Woodbridge, 2002), 373–7, 424–32.
7. Oliver Zimmer, 'Competing Memories of the Nation: Liberal Historians and the Reconstruction of the Swiss Past 1870-1900', *Past & Present* 168 (2000), 194–226.
8. Froissart, *Chronicles*, 88–9; Strickland and Hardy, *The Great Warbow*,114–15, 227–8, 391.
9. Green, *The Hundred Year's War: A People's History*, 136; Strickland and Hardy, *The Great Warbow*, 113–14, 123–4, 134–5.
10. Andrew Ayton, 'Military Service and the Dynamics of Recruitment in Fourteenth-Century England', in *The Soldier Experience in the Fourteenth Century*, ed. Adrian R. Bell, Anne Curry, Adam Chapman, Andy King and Dave Simpkin (Woodbridge, 2011), 31–6.
11. Curry, *Great Battles: Agincourt*, 1–6; 111–40.
12. Holger Riesch, 'Archery in Renaissance Germany', *Journal of the Society of Archer-Antiquaries* 38 (1995), 63–6; Crombie, *Archery and Crossbow Guilds in Medieval Flanders*, 22–6; Potter, *Renaissance France at War*, 70–2, 78–80; Strickland and Hardy, *The Great Warbow*, 118–19.
13. For more on the medieval elements of New Spain see Luis Weckmann, *The Medieval Heritage of Mexico* (New York, 1992).
14. Matthew Restall, *When Montezuma Met Cortés* (New York, 2018), 114, 156–61; Ross Hassig, *Mexico and the Spanish Conquest*, 2nd ed. (Norman, 2006), 58–60, 79–81, 84, 148–9.
15. Bernal Diaz, *The True History of the Conquest of New Spain*, trans. John Ingram Lockhart, 2 vols (London, 1844), 1:248, 1:334; Matthew Restall, *Seven Myths of the Spanish Conquest* (Oxford, 2003), 44–54; Restall, *When Montezuma Met Cortés*, 206–07, 315, 337–8; Hassig, *Mexico and the Spanish Conquest*, 148.
16. Diaz, *The True History of the Conquest of New Spain*, 1:259, 1:350.
17. Restall, *When Montezuma Met Cortés*, 217–21.
18. Diaz, *The True History of the Conquest of New Spain*, 1:57–8, 1:289, 1:334, 1:356.
19. Restall, *When Montezuma Met Cortés*, 166, 171–2, 174–6, 264, 311–13; Hassig, *Mexico and the Spanish Conquest*, 147–9.
20. Diaz, *The True History of the Conquest of New Spain*, 1:57–8.
21. Inga Clendinnen, 'Fierce and Unnatural Cruelty', in *The Cost of Courage in Aztec Society: Essays on Mesoamerican Society and Culture* , ed. Inga Clendinnen (Cambridge, 2010), 71; Hall, *Weapons and Warfare in Renaissance Europe*, 129, 133.
22. Diaz, *The True History of the Conquest of New Spain*, 1:69, 1:71, 1:149, 1:159; Restall, *When Montezuma Met Cortés*, 313–14; Restall, *Seven Myths of the Spanish Conquest*, 143–4.
23. Diaz, *The True History of the Conquest of New Spain*, 1:150–1.
24. Restall, *When Montezuma Met Cortes*, 217; Diaz, *The True History of the Conquest of New Spain*, 2:94; Clendinnen, 'Fierce and Unnatural Cruelty', 69.
25. Matthew Restall and Florine Asselbergs, *Invading Guatemala* (University Park, 2007), 63.
26. Inga Clendinnen, *Ambivalent Conquests: Maya and Spaniard in Yucatán, 1517-1570* (Cambridge, 1989), 15, 31–5.
27. Pedro Sancho, *An Account of the Conquest of Peru*, trans. Philip Ainsworth Means (New York, 1917), 122, 124–5; Restall, *Seven Myths of the Spanish Conquest*, 141.
28. Charles C. Mann, *1491: New Revelations of the Americas Before Columbus* (New York, 2011), 319–25; Gaspar de Carjaval, *The Discovery of the Amazon According to the Account of Friar Gaspar de Carjaval and Other Documents*, ed. and trans. H.C. Heaton (New York, 1934), 56, 100–01, 191–2, 199–200.

29. Gaspar de Carjaval, *The Discovery of the Amazon*, 189–90.
30. Mann, *1491*, 141–51; Restall, *When Montezuma Met Cortés*, 326–30.
31. *A Journal of the First Voyage of Vasco da Gama 1497-1499*, trans. E. G. Ravenstein (New York, 1898), 10–12.
32. Gomes Eanes de Zurara, *The Chronicle of Discovery and Conquest of Guinea*, trans. Charles Raymond Beazley and Edgar Prestage, 2 vols (New York, 1899), 2;161–2.
33. Ibid., 192–3.
34. Michal Tymowski, 'African Perceptions of Europeans in the Early Period of Portuguese Expeditions to West Africa', *Itinerario* 39;2 (2015), 227.
35. Alvise Cadamosto, 'Original Journals of the Voyages of Cada Mosto and Piedro de Cintra to the Coast of Africa, the former in the years 1455 and 1456, and the latter soon afterwards', in R. Kerr, *A General History of Voyages and Travels to the end of the 18th century*, 18 vols (Edinburgh, 1811–24), 2:235, 245.
36. Donald B. Ball, 'Notes on West African Crossbow Technology', *African Diaspora Archaeology Newsletter* 3;1 (1996), 1–5 ; Henry Balfour, 'Origin of West African Crossbows', *Journal of the African Society* 32;8 (1909), 636, 638.
37. Leo Africanus, *The History and Description of Africa*, trans. John Pory and ed. Robert Brown, 3 vols (London, 1896), 3:829.
38. Tymowski, 'African Perceptions of Europeans in the Early Period of Portuguese Expeditions to West Africa', 230; John K. Thornton, *Warfare in Atlantic Africa 1500-1800* (London, 2003), 107–08.
39. Tymowski, 'African Perceptions of Europeans in the Early Period of Portuguese Expeditions to West Africa', 227–8; British Museum Nos. Af1949,46.158 and Af1898,0115.10.
40. Lansiné Kaba, 'Archers, Musketeers, and Mosquitoes: The Moroccan Invasion of the Sudan and the Songhay Resistance (1591-1612)', *The Journal of African History* 22;4 (1981), 471–4.
41. Jens Sensfelder, 'Catalog', in *Die Armbrust Schrecken und Schönheit*, ed. Sven Lüken and Jens Sensfelder (Munich, 2019), 230–2; British Museum No. Af1940,22.2.
42. Balfour, 'Origin of West African Crossbows', 646–9; Blackmore, *Hunting Weapons*, 215; Alm, *European Crossbows*, 11–13.
43. Gomes Eanes de Zurara, *The Chronicle of Discovery and Conquest of Guinea*, 228–9; Balfour, 'Origin of West African Crossbows', 638; Blackmore, *Hunting Weapons*, 215.
44. Balfour, 'Origin of West African Crossbows', 646–9; Blackmore, *Hunting Weapons*, 215; Alm, *European Crossbows*, 11–13.
45. Balfour, 'Origin of West African Crossbows', 642–4; J.R.S. Phillips, *The Medieval Expansion of Europe* (Oxford, 1988), 27–34; Felipe Fernández-Armesto, *Before Columbus: Explorations and Colonization from the Mediterranean to the Atlantic* (Philadelphia, 1987), 188–95.

Conclusion

1. Geoffrey de Charny, 'Livre de Charny', ed. and trans. Nigel Bryant in *The Book of Geoffroi de Charney*, ed. Ian Wilson (Woodbridge, 2021), 68–70, 114–15.
2. *Chivalric Biography of Boucicaut, Jean II Le Meingre*, trans. Craig Taylor and Jane H.M. Taylor (Woodbridge, 2016), 115.
3. Pietro Monte, *Pietro Monte's Collectanea: the Arms, Armour and Fighting Techniques of a Fifteenth-Century Soldier*, trans. Jeffrey L. Forgeng (Woodbridge, 2018), 154.
4. Ibid., 207.

Appendix

1. Payne-Gallwey, *The Crossbow*, 31–7.
2. Ibid., 62–3.
3. Ibid., 90–1.
4. Ibid., 62–5, 90–1.
5. Ibid., 3–10, 31–7, 43–53, 57–65, 73–91, 121–5.
6. Alm, *European Crossbows*, 2–5.
7. Ibid., 11–13.

8. Wegeli's versions were eventually published in Rudolf Wegeli, *Inventar der Waffensammlung des Bernischen Historischen Museums in Bern* (Bern, 1948).
9. Josef Alm, *European Crossbows*, 38.
10. Ibid., 41.
11. Ibid., 64–8, 69–92.
12. For example see: W.F. Paterson, '"Mary Rose" - a Preliminary Report', *Journal of the Society of Archer-Antiquaries* 23 (1980), 29–34.
13. Paterson, *A Guide to the Crossbow*, 7–12.
14. Ibid., 29–30, 38–52, 65–77.
15. Breiding, *A Deadly Art*, vi–vii.
16. Ibid., 30–4, 35–65.
17. MET 29.16.14 and MET 14.25.1575a are not included in the book.
18. Breiding, *A Deadly Art*, 30–1, 37–8; Alm, *European Crossbows*, 64–8.
19. Harmuth, *Die Armbrust*, 152–4.
20. Payne-Gallwey, *The Crossbow*, ix–xii.
21. For example: Harmuth, *Die Armbrust*, 36–7, 138–40.
22. Ibid., 94.
23. Ibid., 125–36.
24. Richter, *Die Hornbogenarmbrust*, 9–19, 170.
25. Ibid., 119–36.
26. See Sensfelder, *Crossbows in the Royal Netherlands Army Museum* 154 for an example of the measurements taken for every crossbow. Ibid., 355–60 for tiller analysis.
27. Ibid., 361–7.

Bibliography

Primary Sources

A Journal of the First Voyage of Vasco da Gama 1497-1499, trans. E.G. Ravenstein (New York, 1898).

Adams, Anthony, and A.G. Rigg, 'A Verse Translation of Abbo of St Germain's *Bella Parisciacae Urbis*', *Journal of Medieval Latin* 14 (2004), 1–68.

Ambroise, *The History of the Holy War*, trans. Marianne Ailes (2003, repr. Woodbridge, 2011).

Andrew of Hungary, *Descriptio Victoriae a Karolo Provinciae comite reportate*, ed. G. Waitz (Hanover, 1882).

'Anglo-Saxon Chronicle', in *The Battle of Hastings: Sources and Interpretations*, ed. and trans. Stephen Morillo (Woodbridge, 1999), 21–7.

Annales Ianuenses, ed. Georg Heinrich Pertz (Hanover, 1862).

Arab Archery, trans. Nabih Amin Faris and Robert Potter Elmer (Princeton, 1945).

Balbi di Correggio, Francisco, *La Verdadera Relacion de Todo Lo Que El Anno De M.D.LXV* (Barcelona, 1568).

______________________, *The Siege of Malta 1565*, trans. Ernle Bradford (London, 1965; repr. 2005).

Cadamosto, Alvise, 'Original Journals of the Voyages of Cada Mosto and Piedro de Cintra to the Coast of Africa, the former in the years 1455 and 1456, and the latter soon afterwards', in R. Kerr, *A General History of Voyages and Travels to the end of the 18th century*, 18 vols (Edinburgh, 1811–24), 2:200–62.

'Carmen de Hastingae Proelio', in *The Battle of Hastings: Sources and Interpretations*, ed. and trans. Stephen Morillo (Woodbridge, 1999), 45–53.

Choniates, Niketas, *O City of Byzantium, Annals of Niketas Choniates*, trans. Harry J. Magoulias (Detroit, 1984).

'Chronica Jutensis', in *Scriptores Minores Historiæ Danicæ*, ed. C.L. Gertz, 2 vols (Copenhagen, 1917), 1:440–70.

Chronicles of the Crusades: being contemporary narratives of the crusade of Richard Coeur de Lion by Richard of Devizes and Geoffrey de Vinsauf; and of the crusade of St Louis, by Lord John de Joinville, trans. J.A. Giles and Thomas Jones (London, 1848).

Chronicles of the First Crusade, ed. and trans. Chris Tyerman (London, 2004, repr. 2012).

Cnutonis Regis Gesta; Sive, Encomium Emmae Reginae, Auctore Monacho Sancti Bertini, ed. Georg Heinrich Pertz (Hanover, 1865).

Comnena, Anna, *The Alexiad*, trans. E.R.A. Sewter (London, 1969).

DeVries, Kelly, and Michael Livingston (eds), *Medieval Warfare: A Reader* (Toronto, 2019).

Diaz, Bernal, *The True History of the Conquest of New Spain*, trans. John Ingram Lockhart, 2 vols (London, 1844).

Eanes de Zurara, Gomes, *The Chronicle of Discovery and Conquest of Guinea*, trans. Charles Raymond Beazley and Edgar Prestage, 2 vols (New York, 1899).

Enguerrand de Monstrelet, *The Chronicles of Enguerrand de Monstrelet*, trans. Thomas Johnes (London, 1840).

Esperandieu, E., *Recueil general des bas-reliefs de La Gaule romaine*, 12 vols (Paris, 1907–28).

Froissart, Jean, *Chronicles*, ed. and trans. Geoffrey Brereton (London, 1968, repr. 1978).

Gaimar, Geoffrey, *LEstorie de Engles*, ed. and trans. Thomas Duffus Hardy and Charles Trice Martin, 2 vols (London, 1889).

Gaspar de Carjaval, *The Discovery of the Amazon According to the Account of Friar Gaspar de Carjaval and Other Documents*, ed. and trans. H.C. Heaton (New York, 1934).

Gaston Phoebus, *Le Chasse de Gaston Phoebus*, ed. and trans. Jean de Foix and Joseph Lavallée (Paris, 1854).

______________, *The Hunting Book*, ed. Gabriel Bise and trans. J. Peter Tallon (London, 1978).

Geoffrey de Charny, 'Livre de Charny', ed. and trans. Nigel Bryant in *The Book of Geoffroi de Charney*, ed. Ian Wilson (Woodbridge, 2021), 53–128.

Geoffrey le Baker, *The Chronicle of Geoffrey le Baker*, trans. David Preest and ed. Richard Barber (Woodbridge, 2012).

'Grandes Chroniques', in *The Battle of Crécy: A Casebook*, ed. Michael Livingston and Kelly DeVries (Liverpool, 2015), 130–5.

Henrici Quinti Regis Anglia Gesta, ed. Benjamin Williams (London, 1850).

Henri de Mondeville, *Chirurgie de Maitre Henri de Mondeville, Chirurgien de Philippe le Bel, Roi de France, Composee de 1306 a 1320*, trans. E. Nicaise (Paris, 1893).

Henry of Huntingdon, *The Chronicle of Henry of Huntingdon*, ed. and trans. Thomas Forrester (London, 1853).

Ibn Shaddād, Bahā' al-Dīn, *The Rare and Excellent History of Saladin*, trans. D.S. Richards (Aldershot, 2002).

Jean le Bel, *The True Chronicles of Jean le Bel*, trans. Nigel Bryant (Woodbridge, 2011).

Jean de Venette, 'Chronicles', in *The Battle of Crécy: A Casebook*, ed. Michael Livingston and Kelly DeVries (Liverpool, 2015), 212–17.

Lawrence of Březová, 'Vavřince Z Březové Kronika Husitská', in *Fontes Rerum Bohemicarum: Prameny Dějin Českých*, ed. Josephus Emler, 8 vols (Prague, 1871–1932), 5:327–534.

Leo Africanus, *The History and Description of Africa*, trans. John Pory and ed. Robert Brown, 3 vols (London, 1896).

Magnus, Olaus, *Historia de Gentibus Septentrionalibus* (Rome, 1555).

Malaspina, Saba, *Die Chronik des Saba Malaspina*, trans. Walter Koller and August Nitschke (Hanover, 1999).

Monte, Pietro, *Pietro Monte's Collectanea: the Arms, Armour and Fighting Techniques of a Fifteenth-Century Soldier*, trans. Jeffrey L. Forgeng (Woodbridge, 2018).

Norwich, Edward, *The Master of Game*, ed. and trans. William A. Baillie-Grohman (New York, 1909).

Orderic Vitalis, *The Ecclesiastical History of England and Normandy*, trans. Thomas Forrester, 4 vols (London, 1853).

Paris, Matthew, *Chronica Majora*, ed. Henry Richards Luard, 7 vols (London, 1876).

Ralph of Coggeshall, *Chronicon Anglicanum*, ed. J. Stevenson (London, 1875).

Richer of Rheims, *Histoire de son temps*, ed. and trans. J. Gaudet, 2 vols (Paris, 1845).

Roger of Howden, *Chronica*, ed. William Stubbs, 4 vols (London, 1868–71).

Roger of Wendover, *Flowers of History*, translated by J.A. Giles, 2 vols (London, 1849).

Sachs, Hans, and Jost Amman, *Das Ständebuch* (Frankfurt am Main, 1568).

Sancho, Pedro, *An Account of the Conquest of Peru*, trans. Philip Ainsworth Means (New York, 1917).

Schiller, Friedrich, *Schiller Volume Three: Joan of Arc, William Tell*, trans. Robert David MacDonald (London, 2005).

Schriber, Hans, *Das Weisse Buch von Sarnen*, ed. and trans. Bruno Meyer (Sarnen, 1984).

Sun Tzu, *The Art of War*, ed. and trans. Ralph D. Sawyer (Boulder, 1994).

Taylor, Craig (ed. and trans.), *Joan of Arc: La Pucelle* (Manchester, 2006).

The Chivalric Biography of Boucicaut, Jean II Le Meingre, trans. Craig Taylor and Jane H.M. Taylor (Woodbridge, 2016).

'The Very Pretty Chronicle of John Žižka The Servant of King Wenceslas', in *John Žižka and the Hussite Revolution*, Frederick G. Heymann (Princeton, 1955).

Tschudi, Aegidius, *Chronicon Helveticum*, ed. and trans. Bernhard Stettler, 22 vols (Basel, 1970–2001).

The History of William Marshal, trans. Nigel Bryant (Woodbridge, 2016).

'Vetus Chronica Sialandie', in *Scriptores Minores Historiæ Danicæ*, ed. C.L. Gertz, 2 vols (Copenhagen, 1918–20), 2:1–72

Villani, Giovanni, 'New Chronicle', in *The Battle of Crécy: A Casebook*, ed. Michael Livingston and Kelly DeVries (Liverpool, 2015), 114–21.

Walsingham, Thomas, *The Chronica Maiora of Thomas Walsingham (1376-1422)*, trans. David Preest (Woodbridge, 2005).

William le Breton, 'Philippide de Guillaume Le Breton', in *Œuvres de Rigord et de Guillaume le Breton*, ed. Henri Delaborde, 2 vols (Paris, 1882–5).

William of Jumièges, 'Gesta Normannorum Ducum', in *The Battle of Hastings: Sources and Interpretations*, ed. and trans. Stephen Morillo (Woodbridge, 1999), 17–19.

William of Malmesbury, *Chronicle of the Kings of England*, trans. J.A. Giles (London, 1847).

William of Poitiers, 'Gesta Willelmi' in *The Battle of Hastings: Sources and Interpretations*, ed. and trans. Stephen Morillo (Woodbridge, 1999), 3–15.

Secondary Sources

Alm, Josef, *European Crossbows: A Survey*, ed. and trans. G.M. Wilson and H. Bartlett Wells (Leeds, 1994, repr. 1998).

Asbridge, Thomas, *The Crusades: The War for the Holy Land* (London, 2010).

______________, *The Greatest Knight* (London, 2015).

Ascani, Enrico and Francesco Gorgo, 'A Technical Study of the Fresco in the 'St Sebastian Chapel', Celle Macra, Italy', *Journal of the Society of Archer-Antiquaries* 42 (1999), 86–8.

Ayton, Andrew, 'Arms, Armour, and Horses', in *Medieval Warfare*, ed. Maurice Keen (Oxford, 1999), 186–208.

Ayton, Andrew, 'The Battle of Crécy: Context and Significance', in *The Battle of Crécy, 1346*, ed. Andrew Ayton and Sir Philip Preston (Woodbridge, 2005), 2–34.

___________, 'The Crécy Campaign', in *The Battle of Crécy, 1346*, ed. Andrew Ayton and Sir Philip Preston (Woodbridge, 2005), 35–107.

___________, 'Crécy and the Chroniclers', in *The Battle of Crécy, 1346*, ed. Andrew Ayton and Sir Philip Preston (Woodbridge, 2005), 287–350.

___________, 'Military Service and the Dynamics of Recruitment in Fourteenth-Century England', in *The Soldier Experience in the Fourteenth Century*, ed. Adrian R. Bell, Anne Curry, Adam Chapman, Andy King and Dave Simpkin (Woodbridge, 2011), 9–59.

___________, and Craig Lambert, 'Navies and Maritime Warfare', in *The Hundred Years War Revisited*, ed. Anne Curry (London, 2019), 169–202.

Bachrach, Bernard S., 'Medieval Siege Warfare: A Reconnaissance', *The Journal of Military History* 58;1 (1994), 119–33.

________________, *Charlemagne's Early Campaigns (768-777): A Diplomatic and Military Analysis* (Leiden, 2013).

Bachrach, David S., *Warfare in Tenth-Century Germany* (Woodbridge, 1998).

_______________, 'Origins of the Crossbow Industry in England', *Journal of Medieval Military History* 2 (2004), 73–88.

_______________, 'Crossbows for the King: The Crossbow During the Reign of John and Henry III of England', *Technology and Culture* 45;1 (2004), 102–19.

_______________, 'Crossbows for the King, Part II: The Crossbow During the Reign of Edward I of England (1272-1307)', *Technology and Culture* 47;1 (2006), 81–90.

_______________, 'The Crossbow in English Warfare from John to Edward I: An Administrative Perspective', *Cithara* 52;1 (2012), 3–21.

Baker, Tim, 'Bow Design and Performance', in *The Traditional Bowyers Bible*, ed. Jim Hamm, 4 vols (Guildford, 1992), 1:43–116.

Balfour, Henry, 'Origin of West African Crossbows', *Journal of the African Society*, 32;8 (1909), 635–50.

Ball, Donald B., 'Notes on West African Crossbow Technology', *African Diaspora Archaeology Newsletter* 3;1 (1996), 1–5.

Baring-Gould, Sabine, *Curious Myths of the Middle Ages* (London, 1877).

Bartlett, Robert, *Blood Royal: Dynastic Politics in Medieval Europe* (Cambridge, 2020).

Benham, Jenny, *Peacemaking in the Middle Ages* (Manchester, 2010).

Bergier, Jean-François, *Wilhelm Tell, Realität und Mythos* (Munich, 1988).

Bernstein, David, 'The Blinding of Harold and the Meaning of the Bayeux Tapestry', *Anglo-Norman Studies* 5 (1982), 40–64.

Blackmore, Howard, *Hunting Weapons: From the Middle Ages to the Twentieth Century* (London, 1971).

Blair, Claude, *European Armour circa 1066 to circa 1700* (London, 1958).

Bloom, Otto, 'Blidemestere, Balistarii og Værkmestere i Kjøbenhavn ca. 1375–1550', *Historisk Tidsskrift* 5;5 (1885), 64–87.

Boffa, Sergio, *Warfare in Medieval Brabant* (Woodbridge, 2004).

Bradbury, Jim, *The Medieval Archer* (Woodbridge, 1985).

___________, *The Medieval Siege* (1992, repr. Woodbridge, 2002).

___________, *The Routledge Companion to Medieval Warfare* (London, 2006).
___________, *The Battle of Hastings* (n.p. 1998; repr. 2010).
Bradford, Ernle, *The Great Siege. Malta 1565* (London, 1961; repr. 2014).
Breiding, Dirk, *A Deadly Art: European Crossbows, 1250-1850* (New York, 2014).
Brown, Robert C., 'Observations on the Berkhamsted Bow', *Journal of the Society of Archer-Antiquaries* 10 (1967), 12–17.
Burns, Robert Ignatius, 'The Medieval Crossbow as Surgical Instrument: an Illustrated Case History', *Bulletin of the New York Academy of Medicine* 48 (1972), 983–9.
Burns, Robert, '100,000 Crossbow Bolts for the Crusader King of Aragon', *Journal of Medieval Military History* 2 (2004), 159–64.
Carr, Mike, *Merchant Crusaders in the Aegean 1291-1352* (Woodbridge, 2015).
Cenni, Alessio, 'Late Medieval Quivers', *Journal of the Society of Archer-Antiquaries* 43 (2000), 27–34.
___________, 'Crossbow-Making in Venice During the 13th Century', *Journal of the Society of Archer-Antiquaries* 47 (2004), 34–43.
Chevedden, Paul E., 'Artillery in Late Antiquity – prelude to the Middle Ages', in *Medieval City under Siege* ed. Ivy A Corfis and Michael Wolfe (Woodbridge, 1999), 131–73.
Christiansen, Eric, *The Northern Crusades* (New York, 1980, repr. 1997).
Clendinnen, Inga, *Ambivalent Conquests: Maya and Spaniard in Yucatán, 1517-1570* (Cambridge, 1989).
___________, 'Fierce and Unnatural Cruelty', in *The Cost of Courage in Aztec Society: Essays on Mesoamerican Society and Culture*, ed. Inga Clendinnen (Cambridge, 2010), 49–90.
Contamine, Philippe, *War in the Middle Ages*, trans. Michael Jones (Oxford, 1984).
Cotterell, Brian, and Johan Kamminga, *Mechanics of Pre-Industrial Design* (Cambridge, 1990).
Credland, Arthur G., 'Crossbow Remains', *Journal of the Society of Archer-Antiquaries* 23 (1980), 12–19.
___________, 'Crossbow Remains (pt. 3)', *Journal of the Society of Archer-Antiquaries* 25 (1982), 16–21.
___________, 'The Crossbow in the Far North', *Journal of the Society of Archer-Antiquaries* 26 (1983), 12–23.
___________, 'Notes on the Crossbow Spanning Bench', *Journal of the Society of Archer-Antiquaries* 33 (1990), 13–22.
___________, 'Crossbow Remains (Part 2)', *Journal of the Society of Archer-Antiquaries* 24 (1991), 9–16.
___________, 'The Crossbow and the Bow in Modern Warfare', *Arms and Armour* 7;1 (2010), 53–103.
___________, 'Aquatic Hunting with the Crossbow', *Jahrblatt der Interessengemeinschaft Historische Armbrust* 5 (2013), 33–49.
Crombie, Laura, 'Representatives of Civic Pride and Cultural Identities; the Ghent Crossbow Competition of 1498', *Arms and Armour* 8;2 (2011), 152–64.
___________, *Archery and Crossbow Guilds in Medieval Flanders, 1300-1500* (Woodbridge, 2018).
Crowley, Roger, *Empires of the Sea: The Final Battle for the Mediterranean 1521-1580* (London, 2008).
Curry, Anne, *Great Battles: Agincourt* (Oxford, 2015).
Curtis Wright, David, 'Nomadic Power, Sedentary Security, and the Crossbow', in John Ferris (ed.), *Calgary Papers in Military and Strategic Studies* 2 (Calgary, 2008), 73–90.
Cushway, Graham, *Edward III and the War at Sea: The English Navy, 1327-1377* (Woodbridge, 2011).
Delle Luche, J.-D., *Des amitiés ciblées: Concours de tir et diplomatie urbaine dans le Saint-Empire, XVe-XVIe siècle* (Turnhout, 2021).
Denny, Mark, 'Bow and Catapult Internal Dynamics', *European Journal of Physics* 24 (2003), 367–78.
___________, *Their Arrows Will Darken the Sun: A Guide to the Evolution and Science of Ballistics* (Baltimore, 2011).
DeVries, Kelly, *Infantry Warfare in the Early Fourteenth century: Discipline, Tactics, and Technology* (Woodbridge, 1996).
___________, 'Catapults Are Not Atomic Bombs: Towards a Redefinition of "Effectiveness" in Premodern Military Technology', *War in History* 4;4 (1997), 454–70.
___________, *Joan of Arc: A Military Leader* (Stroud, 2003).
___________, 'The Use of Chronicles in Recreating Medieval Military History', *Journal of Medieval Military History* 2 (2004), 1–15.

___________, 'The Introduction and Use of the Pavise in the Hundred Years War', *Arms and Armour* 4;2 (2007), 93–100.

___________, 'The Tactics of Crécy', in *The Battle of Crécy: A Casebook*, ed. Michael Livingston and Kelly DeVries (Liverpool, 2015), 447–67.

___________, and Niccolò Capponi, 'The Genoese Crossbowmen at Crécy', in *The Battle of Crécy: A Casebook*, ed. Michael Livingston and Kelly DeVries (Liverpool, 2015), 441–6.

___________, and Robert Douglas Smith, *Medieval Military Technology*, 2nd ed. (Ontario, 2012).

Eddé, Anne-Marie, *Saladin*, trans. Jane Marie Todd (Cambridge, 2011).

Ekdahl, Sven, 'Die Armbrust in Deutschordensland Preussen zu Beggin des 15. Jahrhunderts', *Fasciculi Archaeologiae Historicae* 5 (1992), 17–48.

___________, 'Horses and crossbows: two important warfare advantages of the Teutonic Order in Prussia', in *The Military Orders: Volume 2 Welfare and Warfare*, ed. Helen Nicholson (Aldershot, 1998), 119–51.

Esper, Thomas, 'The Replacement of the Longbow by Firearms in English Army', *Technology and Culture* 6;3 (1965), 382–93.

Fernández-Armesto, Felipe, *Before Columbus: Explorations and Colonization from the Mediterranean to the Atlantic* (Philadelphia, 1987).

Fliegel, Stephen N., 'The Art of War: Thirteenth Century Arms and Armour', in *The Book of Kings: Art, War, and the Morgan Library's Medieval Picture Bible*, ed. William Noel and Daniel Weiss (Baltimore, 2002), 83–97.

Foley, Vernard, George Palmer and Werner Soedel, 'The Crossbow', *Scientific American* 252;1 (1984), 104–10.

Foys, Martin K., 'Pulling the Arrow Out: The Legend of Harold's Death and the Bayeux Tapestry', in Martin K. Foys, Karen Eileen Overby and Dan Terlka (eds), *The Bayeux Tapestry: New Interpretations* (Woodbridge, 2009), 158–75.

France, John, *Western Warfare in the Age of the Crusades 1000-1300* (Ithaca, 1999).

Franken-Stellamans, Erhard, 'A Mathematical Method for Determination of the Appropriate Draw Length for a Given Steel Bow', *Journal of the Society of Archer-Antiquaries* 47 (2004), 92–7.

Frankopan, Peter, *The First Crusade: The Call from the East* (Cambridge, 2012).

Freller, Thomas, *Malta: The Order of St John* (Santa Venera, 2010).

George, Hereford B., 'The Archers at Crecy', *The English Historical Review* 10;40 (1895), 733–8.

Gies, Frances, and Joseph Gies, *Cathedral, Forge, and Waterwheel: Technology and Innovation in the Middle Ages* (New York, 1994).

Gillingham, John, 'The Unromantic Death of Richard I', *Speculum* 54;1 (1979), 18–41.

Gillingham, John, *Richard I* (New Haven, 1999).

Green, David, *The Hundred Years War: A People's History* (Padstow, 2014).

Grummitt, David, *The Calais Garrison: War and Military Service in England, 1436-1558* (Woodbridge, 2008).

Hall, Bert, 'Weapons of War and Late Medieval Cities: Technological Innovation and Tactical Changes', in *Technology and Resource Use in Medieval Europe: Cathedrals, Mills and Mines*, ed. Elizabeth Bradford Smith and Michael Wolfe (Aldershot, 1997), 185–208.

Hall, Bert S., *Weapons and Warfare in Renaissance Europe* (Baltimore, 1997).

Hamm, Jim, 'Sinew Backing', in *The Traditional Bowyers Bible*, ed. Jim Hamm, 4 vols (Guildford, 1992), 1:213–32.

_________, 'Tillering', in *The Traditional Bowyers Bible*, ed. Jim Hamm, 4 vols (Guildford, 1992), 1:257–86.

Hardy, Robert, *The Longbow: A Social and Military History*, 3rd ed. (Fort Worth, 1993).

___________, J. Levy, P. Pratt, Roy King, S. Stanley, B.W. Kooi, D. Adams, and A. Crowley, 'Assessment of the Working Capabilities of the Longbows', in *Weapons of Warre: The Armaments of the Mary Rose*, ed. Alexandra Hildred, 2 vols. (Portsmouth, 2011), 2:622–32.

Harmuth, Egon, *Die Armbrust* (Graz, 1975).

_____________, 'Concerning the One-Foot Crossbow of the High Gothic', *Journal of the Society of Archer-Antiquaries* 28 (1985), 9–12.

Harris, P. Valentine, 'From the Longbow to Crossbow and Back', *Journal of the Society of Archer-Antiquaries* 18 (1975), 9–12.

Hassig, Ross, *Mexico and the Spanish Conquest*, 2nd ed. (Norman, 2006).
Hollister, C. Warren, 'The Strange Death of William Rufus', *Speculum* 48;4 (1973), 637–53.
Holmer, Paul, 'The Military Crossbow in Yorkist England', *Journal of the Society of Archer-Antiquaries* 22 (1979), 11–16.
Holts Booth, Annette, 'Crossbow Production at the Archbishop's Palace, Trondheim, Norway', *Journal of the Society of Archer-Antiquaries* 39 (1996), 94–100.
Hosler, John D., *The Siege of Acre, 1189-1191* (New Haven, 2020).
Hunt, Tony, *The Medieval Surgery* (Woodbridge, 1999).
Jones, Peter N., 'A Short History of the Attack of Armour', *Metallurgist and Materials Technologist* 16;5 (1984), 247–50.
____________, 'The Metallography and Relative Effectiveness of Arrowheads and Armor During the Middle Ages', *Materials Characterization* 29 (1992), 111–17.
____________, and Derek F. Renn, 'The military effectiveness of arrow loops. Some experiments at White Castle', *Château Gaillard* 9-10 (1982), 445–56.
Kaba, Lansiné 'Archers, Musketeers, and Mosquitoes: The Moroccan Invasion of the Sudan and the Songhay Resistance (1591-1612)', *The Journal of African History*, 22;4 (1981), 457–75.
Kaminsky, Howard, *A History of the Hussite Revolution* (Berkeley, 1967).
Karpowicz, Adam, 'Ottoman Bows – An Assessment of Draw Weight, Performance and Tactical Use', *Antiquity*, 81 (2007), 675–85.
Keeble, K. Corey, *European Crossbows in the Royal Ontario Museum* (Toronto, 2008).
Keen, Maurice, *England in the Later Middle Ages* (1973, repr. New York, 2000).
Klopsteg, Paul E., *Turkish Archery and the Composite Bow* (London, 1934).
Kruczek, Jan, *Kusze Katalog Zbiorów* (Pszczyna, 2002).
__________, *Kusze I Ich Twórcy* (Pszczyna, 2013).
Laking, Guy Francis, *A catalogue of the armour and arms in the Armoury of the Knights of St John of Jerusalem, now in the Palace, Valletta, Malta* (London, 1903).
Ławrynowicz, Olgierd, and Piotr A. Nowakowski, 'Hunting Arms and Equipment in Medieval Iconography', *Fasciculi Archaeologiae Historicae* 22 (2009), 85–94.
Li, Xiuzhen Janice, Andrew Bevan, Marcos Martinón-Torres, Thilo Rehren, Wei Cao, Yin Xia and Kun Zhao, 'Crossbows and Imperial Craft Organisation: the Bronze Triggers of China's Terracotta Army', *Antiquity* 88;339 (2014), 126–40.
Lane, Frederic C., 'The Crossbow in the Nautical Revolution of the Middle Ages', *Explorations in Economic History* 7;1-2 (1969), 161–71
Lazzari, Enrico, 'The 15th Century Crossbow in the Burgundy and Italian Iconography: Analysis and Assumptions', *Jahrblatt der Interessengemeinschaft Historische Armbrust* 7 (2015), 95–105.
Liebel, Jean, *Springalds and Great Crossbows*, trans. Juliet Vale (Leeds, 1998).
Loades, Mike, *The Crossbow* (Oxford, 2018).
Mallett, Michael, *Mercenaries and Their Masters* (London, 1975; repr. 2009).
Mann, Charles C., *1491: New Revelations of the Americas Before Columbus* (New York, 2011).
Meier, Dirk, *Seafarers, Merchants and Pirates in the Middle Ages*, trans. Angus McGeoch (Woodbridge, 2006).
Mitchell, Russell, 'Archery *versus* Mail: Experimental Archaeology and the Value of Historical Context', *Journal of Medieval Military History* 4 (2006), 18–28.
__________, 'The Longbow-Crossbow Shootout at Crécy (1346): Has the "Rate of Fire Commonplace" Been Overrated?', in *The Hundred Years War (Part II) Different Vistas*, ed. L.J. Andrew Villalon and Donald J. Kagay (Leiden, 2008), 233–57.
Monteiro, João Gouveia, 'The Battle of Aljubarrota (1385): A reassessment', *Journal of Medieval Military History* 7 (2009), 75–103.
Morris, J.E., 'The Archers at Crecy', *The English Historical Review* 12;47 (1897), 427–36.
__________, *The Welsh Wars of Edward I* (Oxford, 1901; repr. 1996).
Musset, Lucien, *The Bayeux Tapestry*, trans. Richard Rex (Woodbridge, 2005).
Needham, Joseph, Robert D.S. Yates, Krysztof Gawlikowski, Edward McEwen and Wang Ling, *Science and Civilisation in China, Volume 5. Chemistry and Chemical Technology, Part 6. Military Technology: Missiles and Sieges* (Cambridge, 1994).

Nicholson, Edward Williams Byron, *Keltic Researches: Studies in the History and Distribution of the Ancient Goidelic Language and Peoples* (London, 1904).
Nicholson, Helen, *The Knights Hospitaller* (Woodbridge, 2001).
Nowakowski, Andrzej, *Arms and Armour in the Medieval Teutonic Order's State in Prussia*, trans. Maria Abramowicz (Łódź, 1994).
Oakeshott, R. Ewart, *The Archaeology of Weapons Arms and Armour from Prehistory to the Age of Chivalry* (London, 1960; repr. 1996).
O'Connell, Robert L., 'The Life and Hard Times of the Crossbow', *MHQ The Quarterly Journal of Military History* 1;2 (1989), 46–9.
Oman, Charles, *A History of the Art of War: The Middle Ages from the Fourth to the Fourteenth Century* (London, 1898).
Paterson, W.F., '"Mary Rose" - a Preliminary Report', *Journal of the Society of Archer-Antiquaries* 23 (1980).
_________, *A Guide to the Crossbow* (Oldland, 1990).
Payne-Gallwey, Ralph, *The Crossbow: Its Military and Sporting History, Construction and Use* (London, 1903; repr.1995).
Pétrin, Nicole, 'Philological notes on the crossbow and related missile weapons', *Greek, Roman and Byzantine Studies* 33;3 (1992), 265–91.
Phillips, J.R.S., *The Medieval Expansion of Europe* (Oxford, 1988).
Potter, David, *Renaissance France at War: Armies, Culture and Society, c.1480-1560* (Woodbridge, 2008).
Prestwich, Michael, 'Miles in Armis Strenuus: The Knight at War', *Transactions of the Royal Historical Society* Sixth Series, 5 (1995), 201–20.
_______________, *The Three Edwards: War and State in England 1272-1377*, 2nd ed. (New York, 2003).
_______________, *Armies and Warfare in the Middle Ages: The English Experience* (New Haven, 1996; repr. 2006).
_______________, 'The Battle of Crécy', in *The Battle of Crécy, 1346*, ed. Andrew Ayton and Sir Philip Preston (Woodbridge, 2005), 139–57.
Purton, Peter, *A History of the Early Medieval Siege c.450-1200* (Woodbridge, 2009).
__________, *A History of the Late Medieval Siege 1200-1500* (Woodbridge, 2010).
__________, *The Medieval Military Engineer: From the Roman Empire to the Sixteenth Century* (Woodbridge, 2018).
Rackevicius, Gintautas, 'The Crossbow – The Weapon of the Invaders and The Defenders of Vilnius Castle (The Late 14th – Early 15th Centuries)', *Fasciculi Archaeologiae Historicae* 20 (2007), 59–70.
Rau, Carsten, *European Arrowheads and Crossbow Bolts* (Berlin, 2018).
Restall, Matthew, *Seven Myths of the Spanish Conquest* (Oxford, 2003),
_____________, *When Montezuma Met Cortés* (New York, 2018).
_____________, and Florine Asselbergs, *Invading Guatemala* (University Park, 2007).
Richardson, Thom, 'The Introduction of Plate Armour in Medieval Europe', *Royal Armouries Yearbook* 2 (1997), 40–5.
_______________, 'Ballistic Testing of Historical Weapons', *Royal Armouries Yearbook* 3 (1998), 50–2.
_______________, 'Springald Sizes in the 14th Century', in *ICOMAM 50: Papers on Arms and Military History 1957–2007*, ed. Robert Douglas Smith (Leeds, 2007), 326–31.
Richter, Holger, *Die Hornbogenarmbrust* (Ludwigshafen, 2006).
____________, *Die Kunst der Armbrustmacher in Dresden* (Beucha, 2008).
Riesch, Holger, 'Archery in Renaissance Germany', *Journal of the Society of Archer-Antiquaries* 38 (1995), 63–7.
Riley-Smith, Jonathan, *The First Crusade and the Idea of Crusading* (New York, 1986; repr. 2003).
__________________, *The Knights Hospitaller in the Levant c.1070 – 1309* (London, 2012).
Rochholz, E.L., *Tell und Gessler in Sage und Geschichte* (Heilbronn, 1877).
Rogers, Clifford, 'The Military Revolutions of the Hundred Years' War', *The Journal of Military History* 57;2 (1993), 241–78.
____________, 'The Efficacy of the English Longbow: A Reply to Kelly DeVries', *War In History* 5;2 (1998), 233–42.
____________, 'The Age of the Hundred Years War', in *Medieval Warfare*, ed. Maurice Keen (Oxford, 1999), 136–62.

____________, 'The development of the longbow in late medieval England and 'technological determinism'', *Journal of Medieval History* 37;3 (2011), 321–41.

Rogers, R., *Latin Siege Warfare in the Twelfth Century* (New York, 1992).

Rohde, Fritz, 'Concerning the Construction of the Crossbow in the Late Middle Ages', trans. T.M. Hamilton, *Journal of the Society of Archer-Antiquaries* 19 (1976), 17–20.

Runciman, Steven, *The Sicilian Vespers* (New York, 1958; repr. 1960).

Scharf, Gian Paolo G., 'Il <ludus balistarum> nel Medio Evo, con particular riguardo all'area ligure-piemontese', in *Giochi e Giocattoli Nel Medieoevo Piemontese e Ligure*, ed. Rinaldo Comba and Riccardo Rao (Rocca de'Baldi, 2005), 217–24.

Schaufelberger, Walter, *Der Wettkampf in der alten Eidgenossenschaft* (Bern, 1972).

Scott-Mcnab, David, 'Richard Almond: *Medieval Hunting* [Book Review]', *Envoi A Review Journal of Medieval Literature* vol. 11 no. 1 (2002), 23–32.

Sensfelder, Jens, 'Crossbows in the Hunting & Fishing Museum in Munich', *Journal of the Society of Archer-Antiquaries* 43 (2000), 6–11.

Sensfelder, Jens, 'A Gothic Spanning-Belt Hook', *Journal of the Society of Archer-Antiquaries* 43 (2000), 20–4.

____________, 'The "Naked" Cranequin – An Example of the Artistic and Technical Skill in the 16th Century', *Journal of the Society of Archer-Antiquaries* 48 (2005), 23–7.

____________, *Crossbows in the Royal Netherlands Army Museum* (Delft, 2007).

____________, 'Technische Geschichte Der Armbrust am Beispiel Der Berliner Sammlung', in *Die Armbrust Schrecken und Schönheit*, ed. Sven Lüken and Jens Sensfelder (Munich, 2019), 78–89.

____________, 'Catalog', in *Die Armbrust Schrecken und Schönheit*, ed. Sven Lüken and Jens Sensfelder (Munich, 2019), 114–312.

Serdon, Valerie, *Armes Du Diable: Arcs et Arbaletes Au Moyen Age* (Rennes, 2005).

Smets, An, and Baudouin Van Den Abeele, 'Medieval Hunting', in *A Cultural History of Animals in the Medieval Age*, ed. Brigitte Resl (Oxford, 2007), 59–79.

Spiteri, Stephen C., *Armoury of the Knights* (Valletta, 2002).

Stone, John, 'Technology, Society, and the Infantry Revolution of the Fourteenth Century', *Journal of Military History* 68;2 (2004), 361–80.

Strickland, Matthew, and Robert Hardy, *The Great Warbow: From Hastings to the Mary Rose* (Stroud, 2005; repr. 2011).

Thornton, John K., *Warfare in Atlantic Africa 1500-1800* (London, 2003).

Thorpe, Lewis, *The Bayeux Tapestry and the Norman Invasion* (London, 1973).

Tibble, Steve, *The Crusader Armies* (New Haven, 2020).

Tobler, Ludwig, *Schweizerische Volkslieder* (Frauenfeld, 1882).

Tymowski, Michal, 'African Perceptions of Europeans in the Early Period of Portuguese Expeditions to West Africa', *Itinerario* 39;2 (2015), 221–46.

Urban, William, *Last Years of the Teutonic Knights* (Barnsley, 2018).

Vann, Theresa M., 'The Militia of Malta', *The Journal of Medieval Military History* 2 (2004), 137–48.

Vaughan, Richard, *Charles the Bold* (Woodbridge, 2002).

Verbruggen, J.F., *The Art of Warfare in Western Europe during the Middle Ages* (1954, repr. Woodbridge, 1997).

____________, 'Flemish Urban Militias against the French Cavalry Armies in the Fourteenth and Fifteenth Centuries', *Journal of Medieval Military History* 1 (2002), 145–69.

Watson, Keith, 'Estimating Longbow Weights from the Arrow Shaft Data', in *Weapons of Warre: The Armaments of the Mary Rose*, ed. Alexandra Hildred, *Archaeology of the Mary Rose* 3, 2 vols (Portsmouth, 2011), 2:617–18.

Watson, David R., *Iolo's First Book of the Crossbow*, 2nd ed. (1995, repr. Austin, 2009).

Weckmann, Luis, *The Medieval Heritage of Mexico* (New York, 1992).

Wegeli, Rudolf, *Inventar der Waffensammlung des Bernischen Historischen Museums in Bern* (Bern, 1948).

Weiss, Daniel, 'Portraying the Past, Illuminating the Present: The Art of the Morgan Library Picture Bible', in *The Book of Kings: Art, War, and the Morgan Library's Medieval Picture Bible*, ed. William Noel and Daniel Weiss (Baltimore, 2002), 10–35.

White, Lynn, 'Kyeser's "Bellifortis": The First Technological Treatise of the Fifteenth Century', *Technology and Culture* 10;3 (1969), 436–41.

Williams, Alan, *The Knight and the Blast Furnace: a history of the metallurgy of armour in the Middle Ages & the early modern period* (Leiden, 2003).
Wilson, G.M., *Crossbows: Treasures of the Tower* (London, 1975).
Wilson, Guy, 'Pavises in England', *Royal Armouries Yearbook* 2 (1997), 53–4.
__________, 'What's in a Name? One-foot and Two-Foot Crossbows', in *ICOMAM 50: Papers on Arms and Military History 1957–2007*, ed. Robert Douglas Smith (Leeds, 2007), 300–25.
Zimmer, Oliver, 'Competing Memories of the Nation: Liberal Historians and the Reconstruction of the Swiss Past 1870-1900', *Past & Present* 168 (2000), 194–226.

Unpublished Resources

Flynn, Colm, *Crusader Artillery, 1097-1148: Typology, Terminology and Character*, unpublished thesis (2011).
Gorman, Stuart, *Technological Development of the Bow and the Crossbow in Later Medieval Europe*, unpublished thesis (2016).
Pratt-Sturges, Rebekah, *Illuminating the Medieval Hunt: Power and Performance in Gaston Fébus' Le livre de chasse*, unpublished thesis (2017).
Smith, Nicholas James Crosby, *Servicium Debitum and Scutage in Twelfth Century England with Comparisons to the Regno of Southern Italy*, unpublished thesis (2010).
Wilson, G.M., 'Notes on Some Crossbows in the Collection of Glasgow Museum and Art Gallery', unpublished notes (2006).

Online Resources

'Second Lateran Council – 1139 AD', *Papal Encyclicals Online*, accessed 29 May 2021, https://web.archive.org/web/20210529124452/https://www.papalencyclicals.net/councils/ecum10.htm
Ainsworth, Peter, and Godfried Croenen, ed., *The Online Froissart*, accessed 25 July 2021, http://www.dhi.ac.uk/onlinefroissart
Ebert, Jorgan, 'Gamle Danske Sange: Til Refsnæs Kong Valdemar red på jagt.' *ugle.dk*, accessed 25 July 2021, https://web.archive.org/web/20210725212851/https://ugle.dk/til_refsnaes_kong_valdemar.html
Powell-Smith, Anna, *Open Domesday*, accessed 29 May 2021, https://opendomesday.org/

Index